The Myth of Petropower

The Myth of Petropower

Anindya K. Bhattacharya
Adelphi University

Lexington Books
D.C. Heath and Company
Lexington, Massachusetts
Toronto

Library of Congress Cataloging in Publication Data

Bhattacharya, Anindya K
    The myth of petropower.

    Bibliography: p.
    Includes index.
    1. Petroleum industry and trade—Finance. 2. Petroleum industry
and trade—Iran—Finance. 3. Organization of Petroleum Exporting Countries—Finance. I. Title.
HD9560.4.B48   338.2'7'282   76-14665
ISBN 0-669-00735-8

Copyright © 1977 by D.C. Heath and Company.

Published simultaneously in Canada.

Printed in the United States of America.

International Standard Book Number: 0-669-00735-8

Library of Congress Catalog Card Number: 76-14665

To Pilar, Celia, and Andrés

Contents

List of Tables

Preface

This book is a study of the financial dependence of the Organization of Petroleum-Exporting Countries (OPEC) on the Western world with special reference to Iran. It builds upon my article "Financial Realities behind Oil Power in Iran," published in *Foreign Service Journal* (August 1976).

I am grateful to a number of individuals for their invaluable assistance with my research. I would like to thank in particular Assad Homayoun and Youssef Akbar of the Imperial Embassy of Iran in Washington, D.C.; Abbas Ghaffari and F. Varasteh of the Office of the National Iranian Oil Company in New York; Hasan Alaee of the Plan and Budget Organization in Tehran; and Y. Mohammadi of Bank Markazi Iran in Tehran.

I am indebted to a number of graduate students in my courses in international finance and international financial markets for their stimulating comments on the subject. While it is impossible to name them all, special mention must be made of the helpful suggestions offered by John Crofton.

The author alone is responsible for the facts and views contained in this book.

New York   *Anindya K. Bhattacharya*
December 1976

Introduction

A key issue affecting the "interdependence" of nations in this decade and perhaps beyond relates to the twin and interrelated aspects of the economic impact of higher oil prices and the financial problem of "recycling" of petrodollars accumulated in the hands of member states belonging to the Organization of Petroleum-Exporting Countries (OPEC). The future stability or turmoil in the international economic and financial system will be determined to a large extent by how the Western world carries on its relations with major oil-exporting countries in trade and investment areas.

In *Foreign Trade and International Development* (Lexington, Mass.: Lexington Books, D.C. Heath, 1976) I maintained that equity considerations, based on a deliberate redistribution of international income, are much more crucial to an understanding of the viewpoint of nonoil developing countries relating to discriminatory trade intervention than are efficiency criteria of elegant economic logic. The same observation holds with regard to the OPEC ideological position on oil and oil prices—a position characterized by legitimate complaints about the maintenance of "cheap" oil prices by oligopolistic international oil companies and the depletion of OPEC's most "precious," nonrenewable natural resource by "conspicuous" consumption and "inferior" uses of oil in the "squanderous" societies of the industrialized world.[1] The efficiency argument of OPEC—chiefly that world inflation has eroded the purchasing power of oil revenues and that oil prices need to be increased at regular intervals in order to protect the real value of oil income—by contrast is much less convincing. The prices of nonoil goods have not risen in a manner to justify the quintupling of oil prices since late 1973 and the recent increase in oil prices announced at the Doha meeting of OPEC.[2]

"The end of an era" scenario portrayed by many analysts with regard to the international order holds insofar as it refers to the transition from an era of cheap and squandered energy to one of a more responsible use of oil and oil products. The key overall impact of OPEC on international relations lies in focusing attention on inherent limitations to spectacular economic growth in the absence of cheap and abundant sources of oil energy.[3] But the scenario does not hold when it refers to the "power" of OPEC in "bringing down" the Western economic and financial system. Despite the glamor of petrodollars, the fact is that the OPEC economy is still very much a part of the "periphery" of the world economy, closely tied to the industrial "center" in trade and investment areas, not to mention military, technological, and cultural spheres. The so-called power of OPEC is based on the depletion of a single finite primary commodity—a situation that is common to the experience of many nonoil developing countries as well. The financial surplus of OPEC represents the counterpart of the depletion of one natural resource, and the built-in bias of OPEC for higher oil prices translates into a legitimate desire for ensuring regular productive capacities in the future and an understandable concern for survival after the exhaustion of that sole resource.[4]

On equity grounds alone, there is a striking identity of interests between OPEC and nonoil developing countries. Therefore, it is not surprising that, despite the severe economic impact of higher oil prices and increasing signs of the "embourgeoisement" of OPEC, the nonoil developing countries have consistently supported, at least in public forums, the OPEC position on oil prices.[5]

As the year 1977 unfolds, the prospect of an ever-growing oil revenue surplus of OPEC is increasingly being replaced by that of a potential downward trend in such revenues in the years to come—a phenomenon that has aptly been characterized as the "final act in a drama."[6] Dwindling financial surplus of OPEC connotes a twofold significance. First, the concerns of 1974 have become obsolete. The specter of too much money in the wrong hands creating havoc in international trade and finance has lost much of its importance as a major policy issue in the industrialized world. It is now clear that pessimistic analysts exaggerated the chaos scenario and paid little attention to the multiple intricacies of OPEC dependence on the Western world.[7] Second, declining real oil income, coupled with rising commitments, is leading to widening current account deficits of the "high absorbing" members of the OPEC group and to sharp revisions in ambitious development plans of the majority of the cartel members. Of all powerful OPEC members, Iran has consistently adopted the most hardline approach to oil prices and has embarked upon one of the most spectacular development plans. Increasing current account imbalances are having a profound impact on both the Iranian ideology relating to oil prices and the targets of the Iranian Fifth Five-Year Development Plan lasting from 1973-1974 to 1977-1978.

The purpose of this book is to analyze the financial aspects of OPEC oil revenues in terms of both their utilization in domestic development plans and the disposition of OPEC financial surplus abroad. Part I focuses on the basic, two-pronged interdependence of the world and examines the income transfer from the West to OPEC through higher oil prices as well as the recycling of petrodollars back to the West via OPEC investments and imports of goods and services. Part II is a specific study of the Iranian experiment with development planning—an experiment that utilizes oil revenues as the main vehicle for ambitious diversification programs. This part of the study examines the importance of oil income on the Iranian government revenue side and the roles of current and development allocations on the government expenditure side. Part III provides a detailed breakdown of OPEC investments by geographical area as well as by investment instrument and pinpoints the dependence of OPEC nations on the Western world in the area of the investment of surplus oil revenues.

The recycling experience so far suggests that major oil-consuming countries of the West—especially the United States, Japan, and West Germany (which account for nearly half of the industrial world's total output)—could absorb much of the "real" impact of higher oil prices by stepping up exports to fulfill OPEC's insatiable import needs and by opening up additional channels for more diversified investment of OPEC surplus funds. OPEC's surplus funds have been contracting in any case and have been neatly absorbed by the international financial markets with minimum disruption, except in the area of secondary recycling of petrodollars to high-risk borrowers in nonoil developing countries. The financial realities behind oil power, then, illustrate the moral of OPEC dependence on the Western world and suggest a modified continuation, not a total erosion, of the Western system of international trade and finance.

Abbreviations

API	American Petroleum Institute
CD	Certificates of deposit
CIF	Cost, freight, insurance
EEC	European Economic Community
FOB	Freight on board
GATT	General Agreement on Tariffs and Trade
GDP	Gross domestic product
GNP	Gross national product
IGAT	Iran gas trunkline
IMF	International Monetary Fund
LNG	Liquid natural gas
NIOC	National Iranian Oil Company
OAPEC	Organization of Arab Petroleum-Exporting Countries
OECD	Organization for Economic Cooperation and Development
OPEC	Organization of Petroleum-Exporting Countries
OSCI	Oil Service Company of Iran
UAE	United Arab Emirates
UNCTAD	United Nations Conference on Trade and Development

**Part I
Recycling and Interdependence**

1

Western Dependence on OPEC Oil: The Basis of Recycling

At the heart of the recycling issue lies Western dependence on OPEC oil, which in turn has led to a massive transfer of wealth from oil-consuming nations to OPEC. Imports of OPEC oil by industrial countries account for 80 percent of total OPEC export earnings.[1] All oil-consuming countries have historically run large bilateral trade deficits with OPEC, and rising oil prices since late 1973 have accentuated the income transfer problem.

Table 1-1 gives an indication of the huge income transfer involved. It shows that the oil import bill of the United States increased nearly fourfold between 1973 and 1976, from less than $8 billion in 1973 to $34 billion in 1976. Oil import costs of other industrial countries, nearly tripled during the same period as did those of nonoil developing countries. All in all, the non-Communist world transferred some $108 billion in oil import payments to OPEC in 1976 as compared with $36 billion in 1973.

Largely as a result of high oil import costs, Western oil-consuming nations have been running a massive bilateral trade deficit with OPEC. Table 1-2 shows the trade balance positions of Western countries vis-à-vis OPEC for the period 1973-1976. The large increase in the Western deficit in 1974 stands out clearly, as does the reduction in the size of the deficit in 1975—a result of reduced oil imports as well as growing Western exports to OPEC in 1975. In 1976, however, the Western trade balance deficit was higher because of larger oil

Table 1-1
Oil Import Costs, 1973-1976
($ billions)

	1973	1974	1975	1976
United States	7.8	24.6	25.0	33.8
Other Industrial Countries	22.2	61.2	59.0	59.7
of which Japan	6.0	18.9	19.6	22.7
Nonoil Developing Countries	6.0	16.0	17.0	14.4
Total Import Costs	36.0	101.8	101.0	107.9

Source: *International Economic Report of the President*, Transmitted to the Congress in March 1976 (Washington: Government Printing Office, 1976), p. 5; Remarks of Mr. Fujio Matsumuro, Financial Minister of the Embassy of Japan in Washington, D.C., before the Sixty-third National Foreign Trade Convention, New York City, November 16, 1976; and Morgan Guaranty Trust Company, *World Financial Markets* (November 1976), p. 2.

Note: Total figures may not tally with those in individual items due to rounding.

Table 1-2
Western Trade Balance with OPEC, 1973-1975
($ billions)

	1973	1974	1975
United States	-1.0	-12.0	-9.0
European Economic Community	-1.6	-32.0	-18.0
Japan	-4.6	-16.0	-12.0

Source: General Agreement on Tariffs and Trade, *International Trade 1974/75*
(Geneva: General Agreement on Tariffs and Trade, 1975), pp. 98, 105, 107;
and *IMF Survey* (March 15, 1976), p. 91.

imports in the wake of economic recovery and reduced OPEC imports from the West.[2]

Despite the quintupling of OPEC oil prices since late 1973 and despite increasing concern in the West about rising dependence on imported oil, the fact remains that the West continues to be highly dependent on OPEC oil. In spite of national and international commitments to developing alternate sources of energy, oil remains by far the most important component of energy consumption, accounting for nearly half of the global total of primary energy consumption, mainly because it happens to be, even at substantially higher prices, a cheaper and cleaner source of energy.[3] Also, OPEC oil today is in plentiful supply, albeit within the framework of the cartel. OPEC crude oil production today accounts for more than 50 percent of global output, and OPEC proved reserves amount to nearly 65 percent of total proved reserves of the world. Saudi Arabia alone possesses more than one-third of total OPEC proved reserves of crude oil. Table 1-3 shows the geographical distribution of world proved reserves of crude oil.

Dependence on OPEC oil varies greatly within the Western world. Imported oil accounts for more than 90 percent of domestic consumption in Japan, nearly 90 percent for Western Europe, and 40 percent for the United States. As far as overall energy requirements are concerned (including coal, natural gas, crude oil, hydroelectric and nuclear power), the United States relies on imported energy sources for less than 20 percent of total requirements, compared with nearly 90 percent for Japan and 60 percent for Western Europe.[4]

Table 1-4 provides a breakdown of crude oil production and consumption in key geographical areas of the world in 1975. It shows the substantial gap between domestic crude oil production and consumption in the United States (6.2 million barrels per day), Western Europe (12.5 million barrels per day), and Japan (4.8 million barrels per day). The production and consumption figures for Canada and Communist countries are almost evenly balanced, and the small surplus shown in each case is clearly not adequate to meet rising Western demand for imported oil. By contrast, the Middle East has a net export position of

Table 1-3
World Proved Oil Reserves, Yearend 1975[a]

	Billion barrels	Percent of total
Total	662.6	100.0
North America	49.6	7.5
United States	33.0	5.0
Canada	7.1	1.1
Mexico	9.5	1.4
Central and South America	25.9	3.9
Venezuela[b]	17.7	2.7
Ecuador[b]	2.5	0.4
Other	5.7	0.9
Western Europe	25.5	3.8
United Kingdom	16.0	2.4
Norway	7.0	1.1
Other	2.5	0.4
Communist countries	106.9	16.1
U.S.S.R.	80.4	12.1
Other	26.5	4.0
Africa	65.1	9.8
Algeria[b]	7.4	1.1
Libya[b]	26.1	3.9
Nigeria[b]	20.2	3.0
Other	11.4	1.7
Asia-Pacific	21.2	3.2
Indonesia[b]	14.0	2.1
Other	7.2	1.1
Middle East	368.4	55.6
Saudi Arabia[b]	148.6	22.4
Kuwait[b]	68.0	10.3
Iran[b]	65.5	9.9
Iraq[b]	34.3	5.2
UAEAE	29.5	4.5
Qatar[b]	5.9	0.9
Other	16.6	2.5

Source: *International Economic Report of the President*, Transmitted to the Congress in March 1976 (Washington: Government Printing Office, 1976), p. 176.

[a]Estimate.

[b]OPEC member.

18.2 million barrels per day, followed by Africa (3.7 million barrels per day) and South America (1.2 million barrels per day).

The clue to understanding the "energy crisis" of the Western world lies in the fact that, while Western oil consumption has declined slightly since late 1973,

Table 1-4

Crude Oil Production and Consumption, 1975

(million barrels per day)

	Production	*Consumption*	*Net Position*
United States	10.0[a]	16.2	− 6.2
Canada	1.8	1.7	+ 0.1
Western Europe	0.6	13.1	−12.5
Japan	−	4.8	− 4.8
Middle East	19.6	1.4	+18.2
Communist countries	11.6	10.7	+ 0.9
Africa	5.0	1.3	+ 3.7
South America	4.4	3.6	+ 0.8
Australia	0.4	0.6	− 0.2

Source: *International Economic Report of the President,* Transmitted to the Congress in March 1976 (Washington: Government Printing Office, 1976), p. 15.

[a]Includes offshore production.

crude oil production has declined at the same time, thereby virtually canceling out the fall in consumption. Table 1-5 shows oil consumption figures for key Western countries and pinpoints the rather sharp drop in consumption since 1974. However, as Table 1-6 shows, crude oil production has fallen in the United States and Canada during the same period, while it registered a slight increase in Western Europe. At the same time, OPEC crude oil production fell from nearly 55 percent of global total to 50.4 percent in 1975. While crude oil production in Communist countries and non-OPEC Third World countries increased during the same period, the increase has been less than spectacular, to say the least.

In 1976 demand for OPEC oil has grown largely because of increased demand in the wake of the economic recovery in the industrial world and stagnant non-OPEC oil production. Non-OPEC crude oil production is currently about 29 million barrels per day—representing a mere 1-million-barrel-per-day increase over the output level recorded in 1975.[5] Current North Sea oil production is only about 300,000 barrels a day, or 3 percent of Western European oil consumption. The Trans-Alaskan pipeline is expected to bring in only about 1.5 million barrels a day by the middle of 1977—an addition to the current onshore production of 8.4 million barrels per day in the United States.[6]

Even the United States has become more dependent on OPEC oil than ever before. Total oil imports in the United States average more than 7 million barrels a day today, and the OPEC share in this overall import category is nearly 80 percent largely because of the declining share of Canadian oil. Within the OPEC group, the dependence of the United States on Middle East producers has grown steadily over the years, accounting for nearly a third of total imports today. The

Table 1-5
Changes in Western Oil Consumption, 1974-1976
(percent)

	1974	*1975*	*1976 (First Half)*
United States	-3.9	-2.6	2.0
Japan	-3.0	-9.1	6.6
United Kingdom	-5.9	-11.4	-2.5
France	-5.6	-9.0	9.0
West Germany	-10.6	-4.9	7.0
Italy	-0.3	-5.9	3.3

Source: *International Economic Report of the President,* Transmitted to the Congress in March 1976 (Washington: Government Printing Office, 1976), p. 14; and Morgan Guaranty Trust Company, *World Financial Markets* (September 1976), p. 5.

Table 1-6
World Crude Oil Production, 1973-1975
(percent of global total)

		1973	*1974*	*1975*
United States		16.4	15.9	15.8
Canada		3.2	3.0	2.8
Western Europe		0.7	0.7	1.0
Communist countries		17.6	19.2	21.9
OPEC		54.7	54.1	50.4
of which	Venezuela	6.0	5.3	4.4
	Ecuador	0.4	0.3	0.3
	Algeria	1.9	1.7	1.7
	Libya	3.9	2.7	2.9
	Nigeria	3.7	4.0	3.3
	Gabon	0.3	0.3	0.4
	Indonesia	2.4	2.5	2.5
	Saudi Arabia	13.5	15.2	13.3
	Kuwait	5.4	4.6	3.9
	Iran	10.4	10.8	10.1
	Iraq	3.5	3.3	4.2
	UAE	2.3	2.5	2.6
	Qatar	1.0	0.9	0.8
Other		7.6	7.2	8.1
Total		100.0	100.0	100.0

Source: *International Economic Report of the President,* Transmitted to the Congress in March 1976 (Washington: Government Printing Office, 1976), p. 172.
Note: Totals may not add up to 100 due to rounding.

dependence on the Middle East derives largely from the declining share of Venezuelan oil in total imports by the United States.[7] Within the Middle Eastern group of countries, the dependence of the United States on Saudi Arabia continues to grow, with Saudi Arabia currently supplying over 1.2 million barrels a day to the country.[8]

It is estimated that imported oil will account for more than 50 percent of domestic consumption in the United States by 1980, as opposed to more than 40 percent in 1976.[9] The dependence of the United States on imported oil will continue to grow despite the strict oil import control program the country has had since 1955 and despite the slight reduction in the rate of growth of consumption demand for oil.[10] Even if oil consumption demand is cut in half, the country will still be demanding at least 4.5 million barrels a day more in 1985 (21.5 million) than now (17 million).[11]

Table 1-7 shows crude oil production in the United States during 1975-1976. As can be seen, crude oil production in the United States declined by 7.2 percent in 1975 and by 2 percent in 1976. The slight improvement in oil production in 1976 was largely due to new oil discovered offshore, especially off the Santa Barbara coast. However, during the first quarter of 1976, gasoline demand rose by 6 percent over 1975 levels, while heating oil demand rose by 2 percent and residual fuel oil demand stayed at about the same level.[12] On the supply side, due largely to the rising cost of oil drilling relative to controlled prices at home, there is little incentive to find new sources of petroleum or other sources of energy.[13]

Declining crude oil production at home has profound implications for continued Western dependence on OPEC oil in the years to come. The cost and the

Table 1-7
United States Crude Oil Production, 1975-1976
(thousand barrels per day)

Month	1975	1976	
		Amount	*Percent Increase*
January	8590	8239	(4.1)
February	8496	8191	(3.6)
March	9394	8189	(2.4)
April	8375	8160	(2.6)
May	8431	8141	(3.4)
June	8368		
July	8377		
August	8359		
September	8271		
October	8330		
November	8280		
December	8220		

Source: Wilbur L. Gay, "Oil Industry Commentary: An Analytical Service," Goldman Sachs *Investment Research* July 1976, p. 17. Reprinted with permission.

environment-related problems of developing nonoil sources of energy are well known and are beyond the scope of this study.[14] Between 1975 and 1980 demand for oil is expected to grow at an annual average rate of 2 to 3 percent in the United States, 3 to 5 percent in Western Europe, and 5 to 7 percent in Japan. The non-Communist world demand for oil in 1980 is expected to be at least 10 million barrels a day higher than in 1975, and OPEC members are expected to supply more than half of the demand in 1980.[15] Table 1–8 shows the estimated world demand for oil during the 1975-1980 period ranging from a minimum rate of growth of 2.8 percent to a maximum of 5.2 percent, as well as the OPEC supply of oil ranging from a minimum of 24.5 million barrels a day to a maximum of 36.5 million barrels a day in 1980.

It is a reasonable assumption, therefore, that the dependence of the Western world on OPEC oil will continue to grow in the near future. In the absence of another severe recession or a steep oil price rise, it is unlikely that there will be sufficient pressure for a reduction in this dependence originating from either the demand or the supply side.

Table 1-8
Estimated Range of OPEC Crude Oil Production, 1980
(million barrels per day)

	1975	1980 Minimum	1980 Maximum	1975-1980 Growth (%) Minimum	1975-1980 Growth (%) Maximum
Demand					
United States	16.5	18.0	20.0	2	4
Other Western Hemisphere	4.2	5.0	5.5	3	5
Europe	13.0	15.0	16.5	3	5
Japan	5.2	6.5	7.5	5	7
Other Eastern Hemisphere	5.1	6.0	7.0	4	6
Total demand	44.0	50.5	56.5	2.8	5.2
Supply					
Non-OPEC					
United States	10.5	13.0	11.0		
Other Western Hemisphere	3.6	5.0	4.0		
Europe	0.7	5.0	3.0		
Other Eastern Hemisphere	1.7	3.0	2.0		
Total non-OPEC	16.5	26.0	20.0		
OPEC	27.5	24.5	36.5		
Total Supply	44.0	50.5	56.5		

Source: Wilbur L. Gay, "Oil Industry Commentary: An Analytical Service," Goldman Sachs *Investment Research* (July 1976), p. 23. Reprinted with permission.

2

OPEC Financial Surplus: The First
Prong of Recycling

The quintupling of oil prices since late 1973 has resulted in a handsome rise in the financial surplus of OPEC during each year of 1974 to 1976. The current account balance (excluding official transfers) is the most reliable indicator of OPEC's "investable" surplus, although the two figures may not necessarily coincide because of lags in oil receipts and payments.[1]

Table 2-1 shows the overall current account balances of OPEC, industrial countries, nonoil developing countries, and the rest of the world for the 1971-1976 period. As can be seen, the cumulative OPEC current account surplus of more than $140 billion during 1974 to 1976 is offset by aggregate current account deficits in oil-consuming nations of the world, including $20 billion in industrial countries and $100 billion in nonoil developing countries.

In 1974 the OPEC current account surplus increased tenfold over the $6 billion figure recorded in 1973, reflecting the quadrupling of oil prices in late 1973.[2] In 1975, however, the surplus was reduced by nearly half, to $35 billion, because of the interplay of two key factors. Largely as a result of recession-induced conditions in the industrial world, Western oil imports from oil-exporting

Table 2-1
World Current Account Balances,[a] 1971-1976
($ billions)

	Annual Average *1971-1973*	*1974*	*1975*	*1976*	*Cumulative Total* *1971-1973*	*1974-1976*
OPEC	3.4	67.7	34.6	(40)	10.3	142
Industrial countries	16.4	−5.7	21.2	(5)	49.1	20
Southern Europe[b]	0.7	−7.0	−8.0	(−8)	2.2	−23
Nonoil developing countries	−9.9	−29.3	−37.5	(−33)	−29.7	−100
Eastern trading area[c]	−0.5	−3.9	−10.0	(−6)	−1.4	−20

Source: "Prospects for International Trade — Main Conclusions of GATT Study for 1975-76," GATT *Press Release*, GATT 1183 (27 August 1976), p. 28. Reprinted with permission.

[a]Balance of goods, services, and private unrequited transfers, excluding government unrequited transfers.

[b]Includes Greece, Portugal, Spain, Turkey, and Yugoslavia.

[c]Trade balances (exports FOB, imports FOB) taken from foreign trade statistics.

11

countries fell by 10 percent. Total OPEC oil exports declined from 28 million barrels a day in 1974 to 26 million barrels a day in 1975, leading to a fall in OPEC oil export revenues of $5 billion.[3] At the same time, OPEC imports of goods and services continued to rise in 1975, reflecting the desire of OPEC members to diversify and modernize their economies. Total OPEC imports increased in value from $53 billion in 1974 to $87 billion in 1975—an increase of more than two-thirds that provided half of the increase in the nominal value of world trade in 1975.[4]

In 1976, however, the cumulative OPEC current account surplus increased slightly to a sum in the neighborhood of $38 billion, although the size of the surplus is still considerably below that recorded in 1974. The increase of the OPEC financial surplus in 1976 is due to three chief factors: greater consumption demand for oil in the wake of the economic recovery of the West, stagnant non-OPEC oil production, and reduced OPEC import demand as a result of the scaling down of ambitious diversification programs and infrastructural bottlenecks in handling massive amounts of imports. In 1976, the world demand for oil rose by 3.5 percent over 1975, while OPEC commodity imports grew at only 17 to 18 percent compared with the 65 percent growth rate achieved in 1975.[5]

Table 2-2 gives a detailed breakdown of the OPEC current account surplus during 1973-1976 by individual items which enter into the calculation of financial surplus. The figures in this table do not tally exactly with those in Table 2-1 because of the different methods of calculation used (that is, a cash settlements basis versus an accrual basis) and the inclusion or exclusion of certain non-OPEC oil exporters such as Bahrain, Oman, Brunei, Trinidad, and

Table 2-2
OPEC's Current Account
Balance of Payments, 1974-1976
($ billions)[a]

	1974	1975	1976
Oil revenues	102	97	112
Other commodity exports	7	7	8
Services exports[b]	4	5	5
Commodity imports, FOB	36	59	70
Services imports[c]	15	23	30
Investment income receipts	4	6	7
Current account	65	32	33

Source: Morgan Guaranty Trust, *World Financial Markets* (September 1976), p. 4. Reprinted with permission.

[a]Columns may not add due to rounding.

[b]Excluding investment income receipts.

[c]Excluding investment income payments on direct investment.

Tobago.[6] Nonetheless, the overwhelming predominance of oil exports in total OPEC exports stands out clearly, as does the rising size of OPEC imports of goods and services.

The prospective size of OPEC's cumulative financial surplus in 1980 is open to conjecture, and there have been as many projections as there are experts. Estimates of OPEC financial accumulations by the end of the decade have varied widely because of divergent assumptions made about the future price of oil, OPEC export and import volumes, the rate of return on OPEC investments, and other factors.[7] Table 2-3 summarizes the projections made by various experts. As can be seen, the projections vary from a low of $14 billion to a high of $400 billion in constant 1974 dollars.

At the heart of the OPEC financial surplus lie massive oil export revenues, which in turn depend on the pricing policy of OPEC rather than on the market price of oil. The FOB price of crude oil depends on a number of costs: real extraction cost, royalties and taxes paid to the producing country levied as fixed percentages of posted price, and the profit margin of oil companies. The key to the OPEC oil price is the posted price, which is the accounting base upon which royalties and taxes are calculated. Thus, an increase in the posted price of crude oil alone will raise both government revenue and FOB tax-paid cost of all oil.[8] Table 2-4 shows the calculations that enter into the pricing of Persian Gulf crude oil, while Table 2-5 shows the estimated CIF cost of crude oil at Rotterdam for 1975-1976.

Table 2-3
Projections of 1980 OPEC Financial Accumulations
($ billions)

	Current dollars	Constant 1974 dollars
Hollis Chenerey (January 1975)	n.a.	300
Citibank (June 1975)	189	141
Edward Fried (1974)	n.a.	152
Exxon (Spring 1975)	330-380	200-240
Irving Trust Case I (March 1975)	248	158
Irving Trust Case II (March 1975)	22	14
Walter J. Levy (June 1975)	449	286
Mobil (Spring 1975)	303	178
Morgan Guaranty (January 1975)	179	114
OECD (July 1975)	n.a.	215
Thomas Willett et al. (January to May 1975)	n.a.	175-250
World Bank (July 1974)	653	400
World Bank (July 1975)	200-400	120-250

Source: U.S. Treasury, "The Oil Transfer Problem and International Economic Stability," reprinted in *Financial Support Fund*, Hearings before the Committee on Foreign Relations, U.S. Senate, 94th Congress, 1st Session (Washington: Government Printing Office, 1976), p. 157.

Table 2-4
Persian Gulf Crude Oil Prices, 1973-1976
(U.S. dollars per barrel)

	Jan. 1, 1973	Jan. 1, 1974	Jan. 1, 1975	Jan. 1, 1976
Saudi Arabian 34° gravity oil:[a]				
1. Posted price	2.591	11.651	11.251	12.376
2. Royalty (12.5 percent of 1)	0.324	1.456	2.250	2.475
3. Production cost	0.100	0.100	0.120	0.120
4. Profit for tax purposes [1−(2+3)]	2.167	10.095	8.881	9.781
5. Tax (55 percent of 4)	1.192	5.552	7.549	8.314
6. Government revenue (2+5)	1.516	7.008	9.799	10.789
7. Cost of equity oil (3+6)	1.616	7.108	9.919	10.909
8. Cost of participation oil (60 percent of production)	2.330	10.835	10.460	11.510
9. Weighted average cost (7+8)	1.794	9.344	10.240	11.270
10. Weighted government revenue (8−3)	1.694	9.244	10.120	11.150

Source: *International Economic Report of the President,* Transmitted to the Congress in March 1976 (Washington: Government Printing Office, 1976), p. 173.

[a]Saudi Arabian light crude oil 34° API gravity is used as the benchwork for Persian Gulf crude because it is the largest single type of crude oil produced there and represents a good average between higher-priced low-sulfur crude and lower-priced heavier oil.

Table 2-6 shows the operating cost, average government take, and oil company margin relating to Saudi Arabian light crude oil during 1973-1976. This table shows that the operating cost of oil in the Middle East is extremely low (less than $.30) and bears no relationship whatsoever to the posted price of oil. The profit margin of oil companies per barrel declined by 50 percent during the same period. By contrast, the tax-plus-royalty "take" of OPEC governments has increased tenfold between 1973 and 1976. The average government take today accounts for an overwhelming proportion of the posted price of oil and, in effect, represents an excise tax on oil consumers.[9] According to a recent study by Morgan Guaranty Trust Company, the average OPEC government take has more than kept pace with the wholesale dollar prices of manufactured goods.[10]

Despite the reduction in the size of the OPEC financial surplus in 1975 and 1976 relative to 1974, the cumulative surplus to date remains substantial at about $160 billion. However, the distribution of this surplus among OPEC members is acutely uneven, with four countries of the Arabian Peninsula (Saudi

Table 2-5
Estimated Cost of Crude Oil in Rotterdam[a], 1975-1976
($ per barrel)

Month	1975			1976		
	FOB Tax-paid Cost	*Tanker Cost[b]*	*CIF Cost*	*FOB Tax-paid Cost*	*Tanker Cost[b]*	*CIF Cost*
Jan	10.24	1.02	11.26	11.29	1.10	12.39
Feb	10.24	1.04	11.28	11.29	1.10	12.39
Mar	10.24	1.06	11.30	11.29	1.12	12.41
Apr	10.24	1.07	11.31	11.29	1.10	12.39
May	10.24	1.08	11.32	11.29	1.09	12.38
Jun	10.24	1.09	11.33	11.29		
Jul	10.24	1.08	11.32			
Aug	10.24	1.11	11.35			
Sept	10.24	1.09	11.33			
Oct	11.29	1.09	12.36			
Nov	11.29	1.09	12.38			
Dec	11.29	1.07	12.36			

Source: Wilbur L. Gay, "Oil Industry Commentary: An Analytical Service," Goldman Sachs *Investment Research* (July 1976), p. 21. Reprinted with permission.

[a] Average for 34° API light Arabian crude oil.

[b] At AFRA for 200,000-ton vessels. Based on prices prevailing the 15th of the preceding month.

Arabia, Kuwait, Qatar, and United Arab Emirates) accounting for 90 percent of the surplus in 1975 compared with 56 percent in 1974. Saudi Arabia alone, with a population of less than 10 million, accounts for nearly half of the cumulative surplus to date as well as of accumulated net assets abroad.[11] Table 2-7 shows the concentrated nature of the distribution of OPEC financial surplus by member countries during 1974-1976.

The high degree of concentration of OPEC financial surplus in selected countries with a low population base carries profound implications for petrodollar recycling. The relatively low capacity of such Arabian Peninsula countries to absorb surplus funds domestically by way of imports or investments means that much of the petrodollars accumulated in the hands of such countries needs to be placed elsewhere. In 1974-1975 (Saudi Arabian fiscal year 1394-1395) Saudi Arabia spent internally only one-third of its total revenues of $29 billion.[12] The "low absorbers" of the Arabian Peninsula are expected to remain in "structural" surplus for some time to come, although the size of their surplus will shrink by 1980. The high absorbers of the OPEC group (notably Iran, Nigeria, Indonesia, and Algeria), by contrast, either are already in current account

Table 2-6
Oil Company Profit Margin, 1948-1975
($ per barrel of Arab Light)

Date	Posted Price	Sales Price to Affiliate[a]	Average Government Take	Operating Cost	Company Margin[b]
1948	–	–	0.21	0.27	1.56
1951	1.75	1.75	0.58	0.27	0.90
1965	1.80	1.49	0.92	0.10	0.47
1971 – Teheran (Feb.)	2.18	1.85	1.26	0.11	0.48
1972 – Geneva (Jan.)	2.48	1.95	1.44	0.11	0.40
1973 – Geneva (Apr.)	2.74	2.30	1.69	0.13	0.48
1973 – Kuwait (Oct.)	5.12	3.65	3.40	0.13	0.12
1974 (Jan.)	11.65	9.00	9.18[c]	0.17	(0.35)
1974 (Mar.)	11.65	9.50	9.18	0.17	0.15
1975 (Jan.)	11.25	10.46	9.95	0.29	0.22
1975 (Oct.)	12.38	11.51	10.98	0.29	0.24

Source: Testimony of William P. Tavoulareas, President of Mobil Oil Corporation, before the Energy Subcommittee of the Joint Economic Committee of the U.S. Congress, June 2, 1976.

Note: Averaged over equity and buy-back crude from 1973 onward.

[a]This figure would vary by company and transaction.

[b]Government participation and take increased retroactive to January 1.

[c]Before U.S. taxes, if any.

deficit or are moving toward it together with a concomitant negative accumulation of net assets abroad.[13]

The interesting point to note about the OPEC financial surplus is that its size has shrunk and will continue to decline for the rest of the decade, no matter which scenario unfolds. The matter of the disposition of OPEC financial surplus is treated in Chapter 8.

Table 2-7
Distribution of OPEC Current Account Surplus,
1974-1975
($ billions)

	1974	1975
Total	61.5	29.4
Saudi Arabia	22.9	18.0
Iran	10.9	4.7
Kuwait	7.3	4.4
United Arab Emirates	3.1	3.1
Qatar	1.3	0.8
Algeria	0.4	−2.5
Ecuador	0.0	−0.2
Gabon	0.2	0.0
Indonesia	0.4	−2.2
Iraq	2.4	1.0
Libya	1.9	−1.4
Nigeria	4.5	0.7
Venezuela	6.2	3.0

Source: Morgan Guaranty Trust Company, *World Financial Markets* (January 21, 1976), p. 7. Reprinted with permission.

3

Western Exports to OPEC: The Second Prong of Recycling

One of the major ways in which petrodollars have been recycled back to the West is through increasing Western exports to OPEC. The import absorbtive capacity of OPEC countries has turned out to be much greater that what was expected in 1974. Total OPEC imports from the world at large increased from $28 billion in 1973 to more than $48 billion in 1974 and to $75 billion in 1975—increases of 70 percent and 60 percent respectively.[1] In late 1975 and early 1976 OPEC import growth slowed because of both financial and infrastructural constraints, but OPEC is expected to remain a major market for Western products for many years to come.[2]

Table 3-1 shows industrial countries' exports to OPEC during 1974-1975. The countries belonging to the Organization of Economic Cooperation and Development (OECD) increased their exports to OPEC by 77 percent in 1974 and by 64 percent in 1975.

Virtually all OPEC countries (with the exception of United Arab Emirates and Qatar) have embarked upon ambitious development plans using oil revenues for infrastructure or diversification projects. Even without development planning the United Arab Emirates and Qatar are industrializing rapidly and are spending massive sums of money removing infrastructural bottlenecks.

In view of the different levels of progress achieved in various OPEC countries as well as varying amounts of oil revenues actually received, the poorer countries

Table 3-1
Industrial Countries' Exports to OPEC, 1974-1975
(percent change)

	1974	1975
Total OECD	77	64.3
United States	85	63.3
Japan	100	54.8
Germany	80	68.2
United Kingdom	45	77.6
France	64	62.7
Italy	84	66.6

Source: Morgan Guaranty Trust Company, *World Financial Markets* (January 21, 1976), p. 7; (May 1976), p. 3. Reprinted with permission.

of OPEC are concentrating more on basic economic needs such as food, housing, education, employment, and so on, while the more affluent ones are spending a great deal more on high-technology items such as electronics, computers, power tools, and the like.[3] Table 3-2 lists the goods or services that are in demand in various OPEC countries. As one would expect, there is a wide range of requirements for OPEC, and they vary widely by country.

Almost all major OECD countries have benefited from OPEC purchases of goods and services from the West, but the United States, Japan, and West Germany are the chief beneficiaries of increased exports to OPEC. Table 3-3 shows the origin of OPEC imports according to various Western countries. The United States is the leading supplier of goods and services to Iran, Saudi Arabia, and Venezuela—all major countries in OPEC. Japan, by contrast, dominates the markets of Iraq, Kuwait, the United Arab Emirates, and Indonesia, while the United Kingdom is the primary exporter to Nigeria. The ranking, of course,

Table 3-2
Import Needs of OPEC by Importing Countries

Goods or Services	*Countries*
Aircraft	Algeria, Iraq, Kuwait, Venezuela
Agricultural Equipment & Products	Iraq
Apparel	Kuwait
Communication Equipment	UAE, Venezuela, Nigeria
Cooling & Heating Equipment	Iraq, Kuwait
Construction Equipment; Earth Moving Building Materials & Housing	Algeria, Iraq, Ecuador, Nigeria, S. Arabia, UAE
Computers	Algeria
Chemicals, Fertilizers, Pesticides	Iran, Iraq
Electrical Equipment for Industry	Algeria, Venezuela
Electronics	Iran, Iraq, Kuwait
Educational Equipment	Saudi Arabia
Food & Foodstuffs	Algeria, Iran, Iraq, Kuwait, Indonesia, Ecuador, Nigeria
Food Equipment and Seed	Algeria
Fishing Industry Equipment	UAE
Health Care & Medical	Algeria, Venezuela
Metals and Other Raw Materials	Algeria, Venezuela
Mining Equipment	Indonesia
Oil Field Equipment	Iraq, Saudi Arabia, Nigeria
Petroleum Devices	Iran
Power Tools	Iran
Pharmaceuticals	Iraq
Recreational Supplies & Goods	Kuwait
Steel Mills, Pipe	Algeria, Iraq
TV Sets, Appliances, Furniture	Venezuela
Transportation (Road & Rail) Equipment	Algeria
Vehicles & Repair Equipment	Algeria, Iran, Iraq, Kuwait

Source: U.S. Department of Commerce, *Commerce America*, and *Overseas Business Report*, various issues.

Table 3-3
Country Origin of OPEC's Imports in 1975
(percent of total)

United States	18.0
Japan	14.1
Germany	11.4
France	8.2
United Kingdom	7.6
Italy	6.2
Total Six Countries	65.5
Other Industrial Countries[a]	13.8
Developing Countries	10.8
Rest of the World[b]	9.9
Total	100.0

Source: Morgan Guaranty Trust Company, *World Financial Markets* (November 1976), p. 3.

[a]Austria, Belgium, Canada, Denmark, Finland, Greece, Iceland, Ireland, Netherlands, Norway, Portugal, Spain, Sweden, Switzerland, Yugoslavia, Australia, New Zealand, South Africa, and Israel.

[b]Communist countries' exports to OPEC, intra-OPEC trade, and unrecorded military imports by OPEC.

varies in terms of specific orders for specific goods or services. The Japanese, for example, have built more than twenty desalinization plants in the Middle East—a share amounting to about 30 percent of total.[4]

Merchandise exports from the United States to OPEC reached $10.7 billion in 1975, representing over 10 percent of global sales by the United States in that year and twice the share recorded in 1973. Imports from the United States accounted for about one-fifth of total OPEC imports in 1975. OPEC imports of machinery and transport equipment from the United States alone reached $5.7 billion in 1975, almost 13 percent of the exports of capital goods from the United States. OPEC countries accounted for 48.5 percent of the growth in total United States exports in 1975 compared to a mere 5.4 percent in 1973. In addition, military sales to OPEC increased from about $0.2 billion in 1973 to $0.8 billion in 1975, representing some 20 percent of total military sales to foreigners.[5]

In terms of individual markets, Iran ranks as the single largest market among OPEC countries for the United States, absorbing nearly $3.2 billion of exports, followed by Saudi Arabia ($1.5 billion). Under the U.S.-Iran Joint Commission, a target for two-way trade of $40 billion has been set for the period 1975-1980, implying United States exports of over $20 billion. The same export figure—$20 billion—holds for Saudi Arabia as well. The market share of the United States in Saudi Arabia is about 30 percent, followed by about 20 percent in Iran.[6] As for Venezuela, United States exports reached $2.2 billion in 1975, a 27 percent increase over 1974.[7]

Table 3-4 shows the value of United States exports to major OPEC countries during 1973-1976. The dollar value of United States exports should increase in 1976 in all OPEC countries except Iran, Libya, and Algeria. In the future, United States exports will continue to rise, but the rate of growth will probably slow, reflecting curtailment of OPEC development spending. In any case, the commitments of major OPEC countries (Iran, Saudi Arabia, and Venezuela) to high-technology, long-term capital projects such as petrochemicals, gas-gathering systems, and desalinization plants mean booming Western exports to OPEC, especially from the United States, Japan, and Germany.

Table 3-4
United States Exports to Major OPEC Countries, 1973-1976[a]
($ millions)

	1973	*1974*	*1975*	*1976*[b]
Total OPEC	3624	6737	10,700	12,000
of which Iran	769	1730	3200	3000
Iraq	56	285	310	350
Kuwait	117	205	350	490
Saudi Arabia	440	828	1500	2800
UAE	121	229	372	480
Libya	104	139	232	125
Algeria	161	315	600	512
Nigeria	161	285	500	670
Indonesia	441	530	850	1000
Venezuela	1023	1758	2200	2500

Source: U.S. Department of Commerce, *Commerce Today,* and *Commerce America,* various issues.

[a]Figures may not tally due to rounding.

[b]Estimates.

Part II
Financial Realities behind Oil Power in Iran

4 The Iranian Position on Oil Prices

With a population (about 33 million) greater than that of all Persian Gulf countries put together (about 10 million), an industrial base broader than that of other states in the area, and a military machine mightier than that of neighboring countries,[1] Iran—the second largest producer of oil in the OPEC cartel and the fourth largest in the world—has been one of the most influential and articulate members of OPEC, voicing OPEC demands vociferously vis-à-vis the industrialized world.

According to Iranian authorities, the industrial world built its postwar prosperity largely on "cheap oil," the price of which was lowered unilaterally by international oil companies from $2.18 a barrel in 1946 to $1.78 a barrel in the 1960s—a move viewed in Iran as a sheer exploitation of the country's most precious natural resource. Then the prices of imported agricultural and industrial goods that Iran needed rose by an average of at least 300 percent per annum during the same period. The Iranians feel that oil is much too "noble" a product to be put to "indiscriminate inferior" uses such as heating, fuel, and electricity and should be conserved basically for use as feedstock to chemical and petrochemical products—a measure all the more desirable in view of the fact that the supply of finite oil is limited and will probably be depleted by the end of this century. Iran has also emphasized the point that, even with high prices, oil is still cheaper than alternative ways of obtaining energy, and that one can get about 10,000 derivatives from oil in the petrochemical field—derivatives that are unobtainable from other energy sources.[2]

The Iranians feel strongly about the imbalance between proved reserves of fossil fuels and their rate of depletion. Petroleum resources have the least proved reserves, while natural gas and coal have a great deal more. Yet, the former are being depleted at a much faster rate than the latter, with nearly half of the energy consumed in the world being discarded as waste. The Iranians feel that the "low" price of oil is discouraging search for alternative sources of energy, and that the real extraction cost of oil should be set at the minimum price for obtaining shale oil or for liquefaction and gasification of coal at a minimum of $7 or $8 per barrel. From Iran's viewpoint, the key to countering high depletion of oil resources is conservation and efficient energy utilization in the "squanderous" societies of the West.[3]

Using her privileged position in the cartel, Iran has been in the forefront of the OPEC drive to push oil prices upward and has benefited handsomely from the quadrupling of oil prices in late 1973. In July 1973, Iran took complete control over oil production from the Consortium oil companies as opposed to the previous 50 percent control that Iran had acquired since 1953-1954. From here on, the National Iranian Oil Company (NIOC) was empowered, by virtue of the sale and purchase agreement with international oil companies, to decide how much oil should be produced at a given point in time and to sell specified amounts of oil to trading companies as well as to others independently. In effect, a buyer-seller relationship in the oil marketplace was put into effect, as opposed to the former concessionary agreements with international oil companies alone.[4]

The NIOC, a wholly owned organization of the Imperial Government of Iran, is entrusted with a variety of tasks. As a government agency, it determines the production level and the pricing policy of Iranian oil. It is in charge of the operational functions of the oil industry, including exploration, development, production, and marketing of oil. It is responsible for supplying oil to the Iranian domestic market as well as to certain independent markets in the world. Finally, it is also a gigantic nonoil concern that is in charge of Iran's tanker fleet and gas and petrochemical production, among other items.[5]

All oil companies operating in Iran today are wholly owned NIOC subsidiaries, joint venture companies, or major contractors with NIOC. Foreign participation in Iranian oil is limited to either joint venture agreements or contractual agreements. Since July 1974, the Oil Service Company of Iran (OSCI), a private company made up of international oil giants, maintains a 20-year sale and purchase agreement with NIOC permitting OSCI to buy oil at oil wellhead and then export it. The OSCI is owned by British Petroleum (40 percent); Shell (14 percent); Gulf Oil, Mobil, Standard Oil of California, Exxon, and Texaco (7 percent each); Compagnie Française des Petroles (6 percent); and the Iricon Group of Companies (5 percent). The last group consists of American Independent Oil Company, Atlantic Richfield, Getty Oil, Signal Oil and Gas Company, Standard Oil of Ohio, and Continental Oil.[6]

The agreement with OSCI also calls for oil companies to advance annually over the first five years 40 percent of funds required by NIOC for capital outlays in oil and oil-related operations—a payment to be repaid by NIOC in ten annual installments offset against crude oil sales.[7]

As far as oil prices are concerned, the new contract with international oil companies in late 1973 increased the posted price of Iranian oil fourfold, from less than $3 per barrel to nearly $12 per barrel.[8] For each barrel of crude oil exported, the international oil companies pay NIOC: (1) royalty; (2) operating cost, including depreciation; (3) interest payment on 60 percent of capital investment made by NIOC for oil; (4) a "balancing margin" in accordance with participation agreements with other Persian Gulf producers of oil; and (5) income tax on net revenues.[9]

In view of the fact that international oil companies account for nearly 90 percent of Iranian crude oil production and exports, the results of the quadrupling of oil prices have been spectacular for Iran since late 1973.[10] By March

1974 (Iranian year 1352), Iran's oil income went up to nearly $18 billion—an amount 4 times larger than that recorded in the previous year. Iranian crude oil production reached a record 6.5 million barrels per day in 1974, 5.6 million barrels of which were exported—volumes representing twice the size registered in the year before. Iranian GNP recorded a growth rate of 33 percent in 1974, compared with 14.2 percent in 1973. Per capita GNP increased from $556 in 1973 to $815 in 1974.[11]

5

The Pattern of Government Revenues and Expenditures

The Iranian economy is dominated by central planning. The country has had medium-term development plans since 1948.[1] There have been five development plans so far. The Fifth Five-Year Development Plan lasting from 1973-1974 to 1977-1978 (Iranian years 1352-1356) is of the greatest interest, because this latest plan utilizes rising oil income for development purposes. The Fourth Five-Year Plan, launched prior to the quadrupling of oil prices in late 1973, provides a stark contrast to the spectacular Fifth Plan and highlights the increasing importance attached to oil revenues in the Fifth Plan.

The Iranian Fourth Five-Year Development Plan lasting from 1968-1969 to 1972-1973 (Iranian years 1347-1351) indicates the rather modest progress achieved in the country in the absence of significantly higher oil revenues. During the Fourth Plan, Iranian GNP at constant prices rose at an annual average rate of 12 percent, surpassing the target rate of growth of 9.4 percent per annum. Government revenues recorded an average annual rate of increase of 23 percent. Total revenues from the oil sector grew at an average annual rate of 27 percent as compared with 18 percent from all other sources. Oil revenues accounted for more than 50 percent of total government revenue on the average during the lifespan of the Fourth Plan. In terms of current foreign exchange receipts, however, the contribution of the oil sector amounted to 70 percent of total receipts during the Fourth Plan. Oil revenues in dollar terms increased more than twofold, from less than $1 billion in 1967-1968 to more than $2.4 billion in 1972-1973.[2]

In spite of modestly rising oil revenues, the Fourth Five-Year Development Plan was not a complete success. Exports of goods and services increased at an annual average rate of 14 percent as opposed to the planned rate of 15.6 percent. On the other hand, imports rose by 16.3 percent as compared with the planned rate of 13.1 percent. Consequently, net exports grew at only 9.1 percent compared with the target rate of 21.4 percent.[3]

As regards Iranian balance of payments, the net current account was negative each year during the Fourth Plan, while the net capital account was positive throughout, leaving the overall balance of payments in the black only during the last two years of the Fourth Plan. Private capital inflows, more than 50 percent of which came from the United States, was the chief contributing factor in the improvement of the Iranian net capital account during the Fourth Plan.[4]

The Iranian Fifth Five-Year Development Plan is more ambitious than anything undertaken in the country so far. All targets have been revised upward in

29

anticipation of rising oil revenues. The Fifth Plan has committed an outlay of $123 billion, nearly $70 billion of which is allocated to fixed capital investment in the development sector—nearly twice the amount earmarked in the initial 1973 version of the Fifth Plan and 7 times larger than the funds invested in development projects during the Fourth Plan. The Fifth Plan envisages an average annual growth rate of GNP of 25.9 percent in real terms as against the 11.2 percent rate of the Fourth Plan. The contribution of the value-added oil sector in GDP is projected to increase at an annual average rate of 51.5 percent during the Fifth Plan. By the end of the Fifth Plan period (1977-1978), Iranian GNP is expected to reach $55 billion with per capita GNP approaching $1521 in constant 1972 prices.[5]

Table 5-1 summarizes the revenue sources and expenditure allocations during both the Fourth Plan and the Fifth Plan, and shows some interesting features of the Iranian budget. On the revenue side, oil and gas income is expected to account for as much as 83 percent of total revenues during the Fifth Plan compared with 53 percent during the Fourth Plan. The importance of taxes (both direct and indirect) and other revenues pales into insignificance by comparison, accounting for a mere 15 percent of total revenues in the Fifth Plan. Also, because of the excess of expenditures over revenues, the country incurred a deficit of $3.6 billion during the Fourth Plan, one-third of which was financed with foreign loans. The budget deficit is expected to be smaller during the Fifth Plan, but Iran plans on financing the major part of it with foreign loans of $222 million.

On the expenditure side, defense allocations comprise the most important element in the budget, taking up nearly a quarter of total allocations.[6] Table 5-2

Table 5-1
Iran's Revenues, Expenditures, and Deficit Financing, 1968-1978
($ billions)

	Fourth Plan (1968-69/1972-73)	Fifth Plan (1972-73/1977-78)
Revenues	18.5	122.8
Oil and Gas	8.2	98.2
Other	6.7	21.7
Deficit Financing	3.6	2.9
Bank Credit (net)	1.2	—
Sales of Bonds and Bills (net)	1.1	0.7
Foreign Loans (net)	1.3	2.2
Expenditures	18.5	122.8
Current	11.0	50.2
Other	7.5	72.6

Source: Bank Markazi Iran, *Annual Report and Balance Sheet, 1351,* p. 155; *Iran Economic News* (Embassy of Iran), Supplement (March 1975), p. 6.

Table 5-2
Percentage Distribution of Iran's Development Funds, 1968-1978
($ billions and percent)

Sector/Item	Fourth Plan (Actual) %	Fourth Plan (Actual) $	Fifth Plan (Anticipated) %	Fifth Plan (Anticipated) $
Economic Affairs		6.02		44.90
Agriculture and Natural Resources	8.1		6.6	
Industries and Mines	22.3		18.0	
Oil and Gas	11.3		16.8	
Water	8.3		3.6	
Electricity	7.5		6.6	
Transportation and Communications	14.1		10.5	
Telecommunication	7.6		1.9	
Other	0.7		0.5	
Social Affairs		1.49		19.06
Rural Development	1.9		1.3	
Urban Development	1.6		1.6	
Housing	8.2		19.7	
Education	3.5		2.8	
Culture	0.3		0.2	
Health	2.8		1.0	
Social Welfare	1.0		0.1	
Other	0.8		0.6	
Public Affairs		2.50		5.64
Total	100.0	10.01	100.0	69.60

Source: Bank Markazi, *Annual Report and Balance Sheet, 1351,* p. 158; *1353,* p. 34; and Embassy of Iran, *Iran Economic News,* Supplement (March 1975), p. 5.

provides a detailed breakdown of the financial allocation of development funds among the civilian sectors of the economy. This table shows that, during both the Fourth Plan and the Fifth Plan, the largest amount of investment funds was allocated to the economic sector, followed by social affairs and public affairs. However, the percentage distribution of funds shows that the social sector registered the greatest increase since the Fourth Plan. Within the economic sector, the largest amounts have been earmarked for industries and mines (18 percent), followed by oil and gas (16.8 percent) and transportation and communication (10.5 percent)—a clear indication of the government's desire for industrial diversification and for removal of infrastructural bottlenecks. In the social field, housing occupies the highest priority (20 percent), again reflecting the government's awareness of the infrastructural bottlenecks facing Iran today. Housing actually ranks first in the allocation of development funds.[7]

Table 5-3 points to some interesting features of the Iranian balance of payments during both the Fourth and the Fifth Plan periods. It shows that during

Table 5-3
Iran's Foreign Exchange Receipts and Payments, 1968-1978
($ billions)

Item	Fourth Plan (Actual)	Fifth Plan (Projected)
Current Receipts	9.4	114.0
Oil Revenues	7.0	102.0
Other Exports	1.7	4.9
Services	0.7	4.9
Other	–	2.0
Current Payments	10.3	94.7
Imports of Goods	8.5	79.1
Services	1.7	14.3
Other	0.1	1.3
Net Current Account	−0.8	+19.3
Capital Account Receipts	3.0	4.7
Capital Account Payments	1.5	6.5
Net Capital Account	+1.5	−1.8
Overall Balance of Payments	+0.7	+17.5

Source: Bank Markazi Iran, *Annual Report and Balance Sheet, 1351,* p. 170; Embassy of Iran, *Iran Economic News,* Supplement (March 1975), p. 7.

the Fourth Plan the net current account recorded a deficit because of a recurring excess of current payments over current receipts. As mentioned earlier, the net current account was in the red each year during the Fourth Plan. The overall balance of payments improved, however, during the last two years of the Fourth Plan because of the offsetting positive balance on net capital account for each year from 1968-1969 to 1972-1973. The projections for the Fifth Plan, by contrast, show a positive net current account largely because of increasing oil-export receipts and a negative net capital account due to repayments of foreign loans and credits. The overall balance of payments is expected to show a tidy surplus during the Fifth Plan.

The estimates for the Fifth Plan assume, among other things, a sustained rate of growth in oil-export receipts and are subject to revisions depending upon the actual outcome, which, as we know today, has diverged substantially from initial projections.

Changing Oil Revenues and Shifting Investment Priorities

The extremely ambitious targets of the Fifth Development Plan in Iran appear to be headed for some major revisions in the years ahead. The unlimited growth in oil income assumed in the Fifth Plan projections is checked by recession-induced demand conditions in the industrialized world since the last quarter of 1974, although the demand for Iranian oil has picked up again lately.

The impact of declining oil revenues, however, was not felt in Iran until the Iranian year 1354 (1975-1976) largely because of lags in oil-export receipts. In 1352 (1973-1974)—the first year of the implementation of the Fifth Plan—Iran enjoyed an economic boom in the aftermath of the quadrupling of oil prices since late 1973. In that year Iranian GNP increased by nearly 15 percent in constant prices as contrasted with the target rate of growth of 11.4 percent. Oil revenues recorded a rise of 62 percent over the previous year as opposed to a 15 percent increase recorded in 1351 (1972-1973). Per capita GNP increased from $501 in 1351 to $821 in 1352—close to the target of $850. The boom of 1352 resulted in remarkable growth rates in agriculture, the services sector, industries and mines, and domestic money and banking activities. In spite of the large increase in government expenditures, which rose by 39 percent in 1352, government revenues increased even more sharply by 54 percent, thus resulting in a decline of the budgetary deficit. Oil and gas revenues accounted for nearly 67 percent of the increase in total government revenues. The only problem that Iran faced in 1352 was that of domestic inflation brought about both by cost-push pressures and sharp increases in domestic demand.[1]

In 1353 (1974-1975) the Iranian economy continued its boom at a time when the rest of the world faced acute problems of "stagflation." Iranian GDP increased by 13.4 percent in constant prices, and per capita GNP increased from $812 in 1352 to $1344 in 1353 in current prices. The export price of oil accounted for the rapid growth of the Iranian economy, despite a no-growth situation in oil production levels because of the decline in world demand for oil.[2]

In 1354 (1975-1976), however, conditions changed radically for Iran. The recession in industrialized countries, combined with energy conservation measures, led to a sharp decline in world demand for oil, including Iranian oil. Whereas Iranian oil exports had increased by a mere 1.1 percent in constant prices in 1353, they registered a sharp drop of 11.1 percent in 1354. Oil-revenue income in that year fell short of what had been expected in the budget, although the budget had been increased 200 percent to $36 billion.[3]

33

The problem is that, since the last quarter of 1974, petroleum production levels in Iran had been falling, and Iranian oil exports experienced steadily declining volumes—a trend not arrested until March 1976. Much of Iran's problem with oil revenue derives from the fact that the demand for heavy fuel plummeted in the developed world because of depressed conditions in the industrial use of residual fuel and fuel substitution. Western oil companies complain that Iranian heavy crude oil is overpriced in relation both to what the market could take and to other deals available elsewhere due to "price shaving" by certain OPEC members. Also, the oil companies want a renegotiation of the 1973 oil agreement with Iran, whereby they are required to put up 40 percent of total capital costs involved in oil exploration and development of oil fields.[4]

Table 6-1 shows that the contribution of the value added of oil to Iranian GDP at current prices declined from 45 percent in 1974-1975 to 37 percent in 1975-1976, from 84.3 percent of government revenue in 1974-1975 to 77 percent in 1975-1976, and from 89.4 percent of current foreign exchange receipts in 1974-1975 to 87.3 percent in 1975-1976. Table 6-2 shows the rather dramatic decline in both production and exports of Iranian oil during 1352-1354 (1973-1974 to 1975-1976).

At the same time, imports of capital goods, intermediate goods, and consumer goods into Iran have been rising at a much faster rate than anticipated. Some of the increase in the dollar value of imports can be attributed to worldwide inflation, which has also been eroding the real value of Iran's oil income.[5] Table 6-3 shows the dramatic rise in the value of Iranian imports during 1352-1354. As can be seen, during 1974-1975 imports amounted to $6.6 billion—nearly twice the amount recorded in the previous year. In 1975-1976 imports rose to nearly $12 billion—nearly double the amount of the previous year. Tables 6-4 and 6-5 show the composition of Iranian imports by goods and the origin of imports into Iran by major areas of the world. About 53 percent of all

Table 6-1
Contribution of Oil to Iranian Economy, 1973-1976
(percent)

	1973-1974	1974-1975	1975-1976
Contribution of value added of oil to GDP at current prices	30.3	45.0	36.8
Contribution of oil revenue to total government revenue	63.1	84.3	76.7
Contribution of oil sector receipts to total current foreign exchange receipts	81.4	89.4	87.3

Source: Bank Markazi Iran, *Annual Report and Balance Sheet 2534*, p. 7.

Table 6-2
Iran's Production and Exports of Crude Oil, 1973-1976
(percent)

	Production			Exports		
	1973-1974	1974-1975	1975-1976	1973-1974	1974-1975	1975-1976
Total	100.0	100.0	100.0	100.0	100.0	100.0
Oil Service Company of Iran	92.0	92.0	91.2	88.3	86.1	82.9
National Iranian Oil Company	0.3	0.3	0.3	5.8	7.9	12.2
Other Companies Iran-Pan American Oil Company (IPAC)	2.2	2.4	3.4	2.0	2.2	1.9
Irano-Italian Oil Company (SIRIP)	1.2	1.2	0.9	1.2	1.3	0.6
Lavan Oil Company (LAPCO)	3.1	3.2	3.2	2.1	2.0	1.8
Iranian Marine International Oil Company (IMINOCO)	1.2	0.9	1.0	0.6	0.5	0.6

Source: Bank Markazi Iran, *Annual Report and Balance Sheet, 1352, 1353,* and *2534,* pp. 192-93, 194, and 91-92 respectively.

imports are intermediate goods, followed by 30 percent for capital goods and 17 percent for consumer goods. In terms of origin of imports into Iran, the predominance of the OECD countries stands out, accounting for some 85 percent of total imports.

In 1975-1976 the export price of Iranian oil rose by 6.3 percent, while the price of imports rose between 8 and 16 percent. The resulting adverse effect on Iran's terms of trade explains why both Iranian GDP and gross national income rose slowly by only 5 percent and 2.7 percent, respectively, in constant prices.[6]

The impact of declining oil revenues and rising expenditures has been felt most acutely in Iranian balance of payments. Table 6-6 gives a detailed breakdown of Iran's balance of payments during 1352-1354. As can be seen, the decline in oil exports resulted in an increase of current foreign exchange receipts of only 5 percent in 1975-1976. By contrast, current foreign exchange payments increased by 54 percent, reflecting the policy of the country of increasing imports both to keep up with ambitious development projects and to increase domestic supply of goods and services as a counterinflationary measure.[7] As a result, the net current account surplus was reduced from $8.5 billion in 1974-1975 to $2.9 billion in 1975-1976.

At the same time, the net capital account continued to incur a deficit, reflecting government commitments regarding foreign loans, grants, and investments.

Table 6-3
Value of Iran's Imports according to the Standard International Trade Classification, 1975-1976
($ millions)

	Value	Share
Total	11,696	100.0
Foods and Live Animals	1555	13.3
Dairy products and eggs	(88)	0.8
Pulses and their products	(560)	4.8
Sugar, its derivatives, and honey	(537)	4.6
Tea, coffee, chocolate, spices, and other similar products	(36)	0.3
Fruits and vegetables	(131)	1.1
Other	(203)	1.7
Beverages and Tobacco	26	0.2
Raw Nonfood Materials Excluding Fuel Products	369	3.2
Raw caoutchouc	(40)	0.4
Textile yarns not mentioned elsewhere	(196)	1.7
Various unprocessed fertilizers and minerals	(35)	0.3
Other	(98)	0.8
Minerals, Fuel, and Related Products	17	0.1
Vegetable and Animal Oils and Fats	291	2.5
Vegetable Oils	(242)	2.1
Other	(49)	0.4
Chemical Products	835	7.2
Chemicals and their compounds	(140)	1.2
Materials used in dyes and tanning	(69)	0.6
Pharmaceuticals and medical products	(208)	1.8
Plastics, cellulose, and artificial gums	(174)	1.5
Chemical materials and products not mentioned above	(81)	0.7
Other	(162)	1.4
Goods Classified according to Their Primary Materials	3342	28.6
Paper, cardboard, and related products	(142)	1.2
Various yarns and related products	(305)	2.6
Goods made from nonmetal minerals	(242)	2.1
Iron and steel	(1845)	15.8
Other	(808)	6.9
Transportation Machinery and Tools	4973	42.5
Nonelectric machinery	(2539)	21.7
Electric machinery, tools, and appliances	(801)	6.8
Transportation vehicles	(1633)	14.0
Miscellaneous Products	286	2.4
Scientific and professional tools	(168)	1.4
Various products and goods not mentioned above	(72)	0.6
Other	(46)	0.4
Goods Not Classifiable by Type	2	–

Source: Bank Markazi Iran, *Annual Report and Balance Sheet 2534,* pp. 60-61.

Table 6-4
Composition of Iran's Imports, 1975-1976
(percent)

Intermediate Goods	53.1
Capital Goods	29.8
Consumer Goods	17.1
Total	100.0

Source: Bank Markazi Iran, *Annual Report and Balance Sheet 2534*, p. 62.

Table 6-5
Origin of Imports to Iran, 1975-1976
(percent)

OECD Member Countries	84.4
of which United States	19.6
EEC members	39.9
(West Germany)	(17.3)
(U.K)	(8.8)
(France)	(4.4)
(Italy)	(3.6)
Japan	15.9
Socialist Countries	4.8
Others	10.8
of which OPEC	0.9

Source: Bank Markazi Iran, *Annual Report and Balance Sheet 2534*, pp. 62-63.

The deficit on this front amounted to $3.9 billion in 1975–1976 compared with a $3 billion deficit in the previous year. The overall balance of payments incurred a deficit of about $1 billion in 1975–1976 as opposed to a surplus of $5.1 billion in the year before. Iran's external reserves declined from $7.2 billion in 1974–1975 to $6 billion in 1975–1976.

The gap between receipts and payments, combined with the erosion of the real value of oil income, is of great concern to the Iranian authorities. While the Iranian foreign exchange position remains basically healthy, much of it is being drawn down to meet rising commitments abroad. Iran has committed vast sums of money totaling some $12 billion in foreign investment, including $5 billion for France, $3 billion for Italy, $1.2 billion for Britain, $1 billion for Egypt, and $0.6 billion for Pakistan.[8] Recent research indicates that the

Table 6-6
Iran's Balance of Payments, 1973-1974 to 1974-1975
($ millions)

	1973-1974	1974-1975	1975-1976
Net Current Account	353	8529	2946
A. Current receipts	6232	20,922	21,971
1. Receipts of the oil sector	5073	18,672	19,053
Oil revenue from Oil Service Company of Iran	(4490)	(16,216)	(17,296)
Revenue from other oil companies	(182)	(862)	(700)
Purchase of foreign exchange from National Iranian Oil Company[a]	(186)	(1445)	(874)
Purchase of foreign exchange from other oil companies	(215)	(149)	(183)
2. Gas export	87	131	202
3. Purchase of foreign exchange through export of goods	548	563	448
4. Purchase of foreign exchange through export of services	524	1556	2268
Private sector	(326)	(598)	(966)
Public sector	(198)	(958)	(1302)
B. Current payments	−5879	−12,393	−19,025
1. Sale of foreign exchange and utilization of foreign long-term loans and credits for import of goods	−4966	−10,633	−15,924
Private sector	(−2709)	(−5020)	(−7505)
Public sector	(−2257)	(−5613)	(−8419)
2. Gold (other than monetary gold)	−3	−11	−122
3. Sale of foreign exchange for import of services	−910	−1749	−2979
Private sector	(−372)	(−803)	(−1429)
Public sector	(−310)	(−614)	(−1251)
Interest paid	(−228)	(−332)	(−299)
Net Capital Account	625	3254	3636
1. Receipts	1505	702	961
Utilization of foreign long-term loans and credits	(1296)	(257)	(300)
Inflow of foreign private loans and capital	(209)	(445)	(661)
Return of capital from abroad			
2. Payments	−580	−3956	−4600
Repayment of foreign long-term private loans and credits	(−541)	(−1313)	(−729)

Table 6-6 — *(Cont.)*

	1973-1974	1974-1975	1975-1976
Outflow of foreign private loans and capital	(−18)	(−46)	(−259)
Investment abroad	(−1)	(−2388)	(−2941)
Other payments	(−20)	(−175)	(−653)
Grants and aids	(−)ᵃ	(−34)	(−18)
Discrepancies and Currency Rate Adjustments	109	−199	−298
Basic Balance	1387	5076	−991
Nonrecurrent Transfers	−226	−	−
Total Balance	1151	5076	−991
Changes in Foreign Assets	−1151	−5076	991
A. Monetary gold	−	−4	2
B. Foreign exchange holdings (net)	1027	−5062	−1484
Bank Markazi Iran	(−698)	(−5178)	(−1170)
Clearing foreign exchange	(8)	(109)	(45)
Other banks	(−118)	(−20)	(176)
Short-term undertakings	(81)	(37)	(93)
C. Bank Markazi Iran long-term loans	−120	−	−
D. IMF account	−4	−10	−495
Utilization	(0)	(0)	(0)
Redemption	(0)	(0)	(0)
Increased subscription	(0)	(0)	(0)
Grants to the Oil Facility by Bank Markazi Iran	(0)	(0)	(−583)
Special Drawing Rights	(−4)	(−10)	(−12)

Source: Bank Markazi Iran, *Annual Report and Balance Sheet 2534,* pp. 54-55.

ᵃIncludes direct exports and investment revenues.

accumulation of Iranian net assets abroad may have reached its peak already and will probably start to decline, leading to a negative figure by 1980 because of either a running down of accumulated capital assets abroad or extensive borrowing in the private capital markets of the West.[9] In 1975 Iran borrowed more than $265 million from the Eurocurrency markets, and such borrowings exceeded $410 million in the first quarter of 1976 alone.[10]

The decline in oil revenues resulted in sharp cuts in current government expenditures in 1975-1976. Current expenditures had to be reduced for the first time in recent years from 68 percent of total to 64 percent. At the same time, nonoil revenues had to be increased to make up for the shortfall caused by the declining share of oil revenues in total government revenues. However, while the ratio of national savings to GNP has been increasing lately, it is still insufficient for vast investment requirements, and the budgetary shortfall has had to be made up from foreign resources. Tables 6-7 and 6-8 show the actual pattern of government revenues and expenditures during 1352-1354.

In order to get around the problems of dwindling real oil income and rising commitments, the Shahanshah of Iran has also proposed an "indexing" method for valuation of the price of oil, whereby the latter will be tied to the imported price of a "basket" of goods that Iran needs to import. Also, Iranian officials have hinted that the future price of oil will have to continue to go up.

Table 6-7
Revenues of Iranian Government General Budget, 1973-1976
(percent)

	1973-1974	*1974-1975*	*1975-1976*[a]
Taxes	22.0	9.4	12.1
Oil and Gas	52.4	86.5	82.7
Others	22.3	2.5	3.8
Special Revenue	3.3	1.6	1.4
Total	100.0	100.0	100.0

Source: Plan and Budget Organization of the Imperial Government of Iran, *The Budget 1354 and Amended 1353*, Part II, Chart 3.

[a]Projected.

Table 6-8
Iranian Government General Budget Expenditures, 1973-1976
(percent)

	1973-1974	*1974-1975*	*1975-1976*[a]
General Affairs	13.8	8.0	9.9
Defense	24.3	23.3	29.4
Social Affairs	21.4	15.2	17.8
Economic Affairs	22.8	23.5	23.6
Miscellaneous	7.8	8.1	6.2
Repayment of Loans	9.7	6.7	3.2
Investment Abroad	0.2	15.2	9.9
Total	100.0	100.0	100.0

Source: Plan and Budget Organization of the Imperial Government of Iran, *The Budget 1354 and Amended 1353*, Part III, Chart 6.

[a]Projected.

Finally, the Iranian authorities expect that some nonessential imports in the consumer goods sector and the overseas aid and investment programs will have to be reduced.[11]

Lately, however, Iran's crude oil production is on the rise as a result of rising Western demand for oil. Iran's production reached a high of 6.4 million barrels a day in September 1976, compared with 4.9 million in January 1976. Even the demand for Iranian heavy crude, the price of which was reduced by $.07 per barrel, has picked up in recent times. The latest Iranian budget for the year 1355 (1976-1977) expects an increase in revenues of 6 percent, although expenditures will surpass revenues by $2.1 billion, which Iran hopes to finance by possible increases in oil prices or by borrowing.[12]

7

Implications of Iran's Diversification Programs

The overall picture for the Iranian economy is far from bleak, despite some short-run problems arising out of oil-export receipts. The foreign exchange reserve position of the country, amounting to more than $6 billion at the end of March 1976, remains basically healthy. The domestic inflation rate, which had been averaging at least 20 percent per annum during 1973-1974 and 1974-1975, has been controlled effectively by tight domestic monetary and credit policies, official subsidies of basic commodities, and severe crackdowns on black marketeers.[1]

Even assuming the worst scenario for Iran in the future—namely, a drop in demand for oil combined with a glut in the supply of oil—the vast diversification programs already undertaken will have borne fruit by the 1980s. By around 1985 to 1990 petrochemical production, together with natural gas, iron and steel, copper and rubber, and invisible exports such as shipping and oil tankers, will hopefully bring in sufficient revenues to make up for any possible loss in oil income.[2] It is estimated that, with a 5 to 10 percent utilization of petroleum as petroleum feedstock, Iran will be able to earn an income that will at least equal that earned on crude oil exports.[3] As for natural gas, Iran possesses the second largest reserve (after the Soviet Union) of such gas in the world and expects to spend a substantial amount during the Fifth Plan developing gas-related industries.

The Iranian experiment with natural gas started in 1967 with the founding of the National Iranian Gas Company as a wholly owned subsidiary of NIOC. Iran has allocated $2.5 billion during the Fifth Plan developing the natural gas sector, which has an estimated production capacity of 5.8 billion cubic feet in 1980.[4] In 1970 the Iran Gas Trunkline (IGAT) was initiated to export natural gas to the Soviet Union and serve Iranian cities at the same time. By the end of 1974, this project had exported more than 1.1 billion cubic feet of natural gas to the Soviet Union. In addition, Iran has entered into joint venture agreements for the export of gas and liquid natural gas (LNG) to the United States, France, West Germany, and Austria.[5]

Of all countries in the world, Iran has the greatest gas exporting potential. The United States is the world's largest natural gas producer, but it imports a great deal of natural gas from Canada and Mexico. The Soviet Union ranks second in terms of production but has the largest gas reserve in the world. But, due to considerable domestic consumption, it is not expected to become a major exporter in the near future. Table 7-1 lists the potential exporters of gas as well as the main consuming countries of gas in the world.

Table 7-1
Natural Gas Reserve/Production Ratio, 1974
(Trillion cubic feet)

Potential Exporters	End 1974 Estimated Reserves	1974 Production	R/P Ratio (Years)
U.S.S.R.	632.0	9.2	69
Iran	398.0	1.8	221
UAE	100.0	0.4	250
Saudi-Arabia	96.0	1:2	80
Algeria	92.3	0.7	132
Netherlands	85.0	2.9	29
Canada	68.0	3.3	20
Nigeria	45.0	0.6	75
Indonesia	34.7	0.2	173
Afghanistan	4.9	0.2	24
Norway	27.4	0.4	68
Libya	28.0	0.5	56
Brunei	20.0	0.3	66

Main Consuming Countries	End 1974 Estimated Reserves	1974 Production	R/P Ratio (Years)
U.S.	270.0	22.3	12
U.S.S.R.	632.0	9.2	69
Eastern Europe	24.2	2.0	12
West Germany	12.4	0.7	18
U.K.	42.0	1.2	35
France	6.6	0.3	22
Italy	6.1	0.6	10
Other West Europe (Excl. Netherlands)	29.3	0.5	59
Japan	1.2	0.1	12

Source: *Iran Oil Journal*, No. 189 (Spring 1976), pp. 30–31.

As for petrochemicals, Iran plans to invest $8 billion in this sector during the Fifth Plan. The Iranians expect that petrochemicals, which presently account for about 10 percent of energy consumption in the world, will provide nearly half of total world energy consumption by the end of this decade. The National Iranian Petrochemical Company was founded in 1965 as a wholly owned subsidiary of NIOC, and it has engaged in seven joint ventures so far, including two with Mitsui and Mitsubishi of Japan, to be completed around 1976 to 1978. The four major petrochemical complexes of Iran—namely, Shahpur Chemical Company, Iran Chemical Fertilizer Company, Abadan Petrochemical Company, and Khark Chemical Company—have already been producing close to capacity, and exports of such products amounted to slightly less than $200 in 1975.[6]

It is interesting to note that the petrochemical sector in Iran also attracts the largest amount of foreign private capital. Table 7-2 shows the inflow of foreign private capital into Iran by type of activity according to countries.

The diversificiation plans of Iran are geared to escaping from excessive reliance on finite oil resources and replacing oil revenues by other sources of income. Oil revenues, which presently account for between 75 and 80 percent of the government budget, are expected to decline to about 90 percent by 1988. Industrial growth of the nonoil sector of the economy is, therefore, of crucial importance to Iran's diversification strategies.[7]

In the overall relationships between policymakers of OPEC and those of industrialized countries, the implications of the Iranian experiment in diversification programs should not go unnoticed. By the end of the Fifth Plan period

Table 7-2
Private Foreign Investment in Iran by Type of Activity and Countries, 1975-1976
(percent)

Activity	
Agroindustry	6.6
Mining	0.3
Food	0.3
Rubber	16.4
Pharmaceuticals and chemicals	6.4
Petrochemicals	22.9
Metallurgical	6.9
Electrical and electronic industries	6.3
Automobile industry and transportation	17.0
Building materials and construction	4.2
Hotels	0.4
Other	12.3
Total	100.0
From	
United States	14.8
United Kingdom	3.4
Germany	6.0
France	15.8
Japan	42.9
Others[a]	17.1
Total	100.00

Source: Adapted from: Bank Markazi Iran, *Annual Report and Balance Sheet 2534*, p. 89.

[a]Includes mixed companies.

(1977–1978), Iran is expected to import nearly \$93 billion worth of goods and services from foreign countries, thereby opening up unlimited opportunities for exports from the Western world. The United States ranks first as the biggest supplier of imports into Iran, and American exports to Iran have been doubling in value each year since 1973. Today, Iran ranks among the fifteen largest buyers of American goods and services, and the American share in total Iranian imports is more than 20 percent.[8]

The import factor implied in Iran's ambitious diversification plans, combined with Iranian dependence on private financial markets to raise the needed loans to offset adverse balance-of-payments movements, makes Iran an attractive proposition for the international business and financial community.

Part III
The Disposition of the OPEC Financial Surplus

8

OPEC Investments Abroad

The financial surplus of OPEC has been invested in a variety of ways and is indicative of OPEC dependence on the Western world. The vast proportion of this surplus has been recycled to the West via international money and capital markets.

Table 8-1 shows the disposition of OPEC surplus from 1974 through the first half of 1976. It shows both the changing geographical concentration and the changing investment strategies of OPEC surplus during the time in question. Thus, while in 1974 the Eurocurrency markets absorbed 37 percent of total surplus, their share fell to 24 percent in 1975 and to one-quarter in the first half of 1976. Similarly, sterling investments in Britain fell from 11 percent of total to almost nothing in 1975 and to net disinvestment in the first half of 1976. The share of petrofunds invested in the United States fell from 22 percent in 1974 to 16 percent in 1975, but increased to nearly 44 percent in the first half of 1976. Also, there has been a marked increase in OPEC investments in other areas, notably in direct loans and grants to governments and international organizations as well as in areas such as Eurobonds, direct investments, and local currency deposits in countries other than the United Kingdom and the United States. Contributions to international organizations rose from 8 percent in 1974 to 10 percent in 1975, while bilateral aid to nonoil developing countries increased from 5 percent in 1974 to 14 percent in 1975.[1]

As far as investment instruments are concerned, there has been a marked shift from short-term, money market instruments to longer-term, capital market instruments. Thus, investments in bank certificates of deposit and Treasury bills declined from 66 percent of total in 1974 to 24 percent in the first half of 1976. By contrast, investments in bonds and notes as well as in direct loans to developed countries rose from 14 percent in 1974 to 30 percent in the first half of 1976. Equity investments recorded a similar increase from 2 percent in 1974 to 9 percent in the first half of 1976.[2] Net new investments in bank deposits and Treasury bills in the United States and United Kingdom fell from 27 percent in 1974 to 5 percent in 1975. By contrast, net new bond and equity investments in the United States rose from $1.6 billion, or 3 percent in 1974 to $4.7 billion, or 16 percent, in 1975.

The shift from short-term, liquid assets in 1974 to longer-term commitments in 1975–1976 reflects the more favorable interest rate movements in long-term debt instruments. After hitting an all-time high in 1974 and in early 1975, short-term interest rates fell sharply in major money markets of the world, necessitating a change of OPEC strategy.

Table 8-1

The OPEC Surplus and Its Disposition, 1974-1976

($ billions)

	1974	1975	1976 First Half
Financial surplus[a]	55.0	31.7	14.9
Investments in the United States	12.0	10.0	6.5
Bank deposits and treasury bills	9.3	1.1	1.6
Bonds and notes	1.1	3.6	3.1
Equities	0.4	1.6	1.1
Other[b]	1.1	3.7	0.7[c]
Investments in the United Kingdom	7.2	0.2	− 0.8
Bank deposits (in £)	1.7	0.2	− 0.9
Treasury bills (in £)	2.7	− 0.9	− 0.8
Government bonds (in £)	0.9	0.4	0.1
Equities and property (in £)	0.7	0.3	0.2
Direct loans (foreign currency)	1.2	0.2	0.6
Eurocurrency bank deposits, plus domestic currency deposits in countries other than the United States and the United Kingdom	22.7	9.1	3.7
International organizations	4.0	2.9	1.8
IMF oil facility	1.9	2.7	1.3
World Bank and other regional development institutions	2.1	0.2	0.5
Grants and loans to developing countries	2.5	4.0	2.5
Direct loans to developed countries other than U.S. and U.K.	4.5	2.0	0.7
Other net capital flows[d]	2.1	3.5	0.5

Source: Morgan Guaranty Trust Company, *World Financial Markets* (September 1976), p. 6. Reprinted with permission.

[a]Current account surplus adjusted for lag between oil exports and payments

[b]Includes real estate and other direct investment in the United States, prepayments for United States exports, debt amortization, and other items.

[c]Includes a shift of $1 billion from direct investment in the United States into banking and portfolio assets reflecting the payment of dividends to a Middle Eastern country on its participation in a U.S.-incorporated petroleum company; in the past, dividend accruals had been treated as a build-up of OPEC direct investment in the United States.

[d]Includes investments in Eurobonds, other portfolio and direct investments, and debt repayments.

OPEC Investments in the Eurocurrency Markets

In 1974 the Eurocurrency markets were saturated with petrodollars. Gross inflow of new OPEC funds into the Euromarkets amounted to $24 billion, or 37 percent of total OPEC surplus. Claims on OPEC countries by Eurobanks amounted to a mere $1 billion, leaving a net flow of $23 billion. The liabilities of banks monitored by the Bank for International Settlements (BIS) to the Middle East OPEC countries alone rose by nearly $20 billion, an increase of 200 percent in 1974 over 1973. In 1975, however, largely as a result of declining rates in Eurocurrency rates, OPEC deposits slowed considerably. In 1975 direct deposits by OPEC increased by $5.5 billion compared with over $20 billion in 1974.[3] Table 8-2 shows the net borrowing position (that is, the difference between liabilities and assets) of BIS banks vis-à-vis the Middle East during 1973-1975, while Table 8-3 shows the cumulative net lending position of BIS banks vis-à-vis OPEC in 1975 by individual countries. As can be seen from Table 8-3, the only major net borrower from the Eurocurrency markets is Indonesia, with net debts of about $2 billion. Other borrowers in the OPEC group include Ecuador and Gabon. Also, the contrast between the high and low absorbers of the Middle East stands out, with the high absorbers—particularly Iran, Libya, and Algeria—drawing their Eurobalances for import payments. In 1975 OPEC as a group borrowed some $1.8 billion in new Eurofunds.[4]

The pattern of the net borrowing position of Eurobanks vis-à-vis OPEC is particularly pronounced in the case of United Kingdom banks, which control more than half of the market share. The net foreign currency borrowing position of U.K. banks vis-à-vis OPEC increased from $2.9 billion at the end of 1973 to $16.2 billion at the end of 1974 and $19.3 billion at the end of 1975.[5] Table 8-4 shows that the vast proportion of the total net borrowing position of U.K. banks is concentrated in the Middle East, followed by Venezuela and Ecuador. By contrast, U.K. banks possessed net claims on Nigeria, Gabon, Indonesia, and Algeria.

Table 8-2
Net Borrowing Position of BIS Banks vis-à-vis the Middle East, 1973-1975
($ millions)

	1973	*1974*	*1975*	*1976 (March)*
Liabilities	10,260	29,830	33,176	51,200
Assets	2550	3570	4339	14,700
Net Borrowing (−)	−7710	−26,260	−28,837	−36,500

Source: The Bank for International Settlements, *Forty-fifth Annual Report*, April 1974-31 March 1975 (Basel, 9 June 1975), p. 140; *Forty-sixth Annual Report*, April 1975-March 1976 (Basel, June 1976), p. 87; and *IMF Survey* (August 16, 1976), p. 247.

Table 8–3
Net Lending (+)/Net Borrowing (–) Position of BIS Banks vis-à-vis OPEC, 1975
($ millions)

	Liabilities	Assets	Net Position
Total OPEC	50,612	13,641	−36,971
Ecuador	163	262	+ 99
Venezuela	7252	2964	− 4288
Gabon	69	144	+ 75
Nigeria	381	270	− 111
Indonesia	473	2478	+ 2005
Middle East	33,176	4339	−28,837
Low absorbers[a]	24,570	1163	−23,407
High absorbers[b]	8606	3176	− 5430
Algeria	1285	1383	+ 98

Source: Bank for International Settlements, *Forth-sixth Annual Report*, April 1975–March 1976 (Basel, June 1976), pp. 86–87.

[a]Kuwait, Qatar, Saudi Arabia, and United Arab Emirates.

[b]Bahrain, Iran, Iraq, Libya, and Oman.

Table 8–4
Net Liabilities (–) and Net Claims (+) in
Foreign Currencies of United Kingdom
Banks, December 31, 1975
($ millions)

Venezuela	−1057
Ecuador	− 14
Nigeria	+ 12
Gabon	+ 25
Indonesia	+ 197
Algeria	+ 101
Middle East[a]	−8665

Source: Bank of England, *Quarterly Bulletin*, Vol. 16, No. 1 (March 1976), Table 21/2.

[a]Includes Bahrain, Iran, Iraq, Kuwait, Libya, Oman, Qatar, Saudi Arabia, and the United Arab Emirates.

In 1974 the highest percentage of OPEC surplus (37 percent of the total, or $23 billion) was placed in the Euromarkets, compared with 30 percent in domestic markets of non-OPEC countries, 7 percent in international agencies, and 26 percent in direct government lending, aid, and grants. The reasons for the high concentration of petrodollars in Euromarkets in 1974 were twofold. First, petrofunds faced various restrictions in direct investments in domestic markets of the West relating to transfer of capital, profits, royalties, interests, service fees, and purchase and sale of equity interest in a company. Chief among such restrictions were the 30 percent withholding tax in the United States on interest and dividends paid to foreign investors and the negative interest rate on local currency deposits in Germany.[6] Second, the safety and the secrecy of Eurodeposits, combined with high returns on short-term investments in the Euromarkets, coincided with OPEC preference for liquidity and short-term maturities.[7] The result was that an overwhelming proportion of petrodollars was kept in very short-term maturities (ranging from call to 3 months) in the Eurocurrency markets.

The marked slowdown in Eurocurrency deposits by OPEC in 1975 reflected both declining OPEC revenues and interest-rate differential. As noted earlier, aggregate OPEC surplus declined from roughly $65 billion in 1974 to $35 billion in 1975. At the same time, interest rates in the Euromarkets fell sharply beginning in 1975, reflecting the impact of vast petrodollar deposits in the market and the difficulty in finding creditworthy borrowers to absorb new injections of funds. Table 8-5 pinpoints the sharp drop in interest rates in the Eurocurrency markets for the period 1973–1976.

In early 1974 OPEC's main concern in the Euromarkets was obtaining the highest possible return in the shortest possible maturity period. Since then there has been a marked evolution in the pattern of OPEC involvement in the Eurocurrency markets. Because of the recent problems of Euromarkets—declining short-term rates, concern about capital ratios of banks, and talks about possible refinancing of external debt of some nonoil developing countries—much less OPEC money is flowing into Euromarkets now, although some OPEC countries, particularly Saudi Arabia and Venezuela, continue to build up massive Eurodeposits. At the same time, OPEC members are getting increasingly involved in Eurocredit syndications as well as in Eurobond placements and syndications.[8] Particularly noteworthy is the increasing involvement of Arab banks and financial institutions in loan syndications and bond placements. A number of Arab investment banks are playing an increasingly important role as lead managers or co-managers in such activities, and some Euroloans and bond issues have been denominated in Arab currencies, particularly in Kuwaiti dinar. Chief among the

Table 8-5
Eurodollar Deposit Rates, 1973–1976
(Prime banks' bid rates in London in percentages)

	1973 (Dec.)	1974 (Dec.)	1975 (Dec.)	1976 (Sept.)
Overnight	9.75	8.00	5.13	5.08
7-day fixed	9.50	9.38	5.19	5.31
1 month	10.06	9.75	5.38	5.44
3 months	10.13	10.19	5.81	5.75
6 months	10.13	10.19	6.63	6.13
12 months	9.56	9.75	7.19	6.44

Source: Morgan Guaranty Trust Company, *World Financial Markets* (October 1976), p. 15. Reprinted with permission.

Arab banks playing such a role are Banque Arabe et Internationale d'Investissement, Kuwait Investment Company, Union de Banques Arabes et Françaises, Banque Intercontinentale Arabe, and Banco Arabe Espanol, Arab Bank Limited.[9]

The increasing attraction of the Eurobond market for OPEC investors reflects both the more favorable spread in longer-term commitments and the availability of quality borrowers (e.g., multinational corporations and governments) in the bond market. In 1974 OPEC provided $3.2 billion to the foreign and international bond markets, or 27 percent of total transactions, of which $674 million was Eurobond issues. Of the total amount, $2.4 billion was lent to development institutions (with the World Bank alone taking in $2.2 billion). In 1975 OPEC supplied about $1 billion, $895 of which was Eurobond issues, many of them denominated in Kuwaiti dinar and Saudi riyal.[10] In the absence of another major turnaround in interest rate structures (such as what happened in 1974), OPEC involvement in the Eurobond market will continue to grow in the future.

OPEC involvement in the Eurocurrency markets is not limited, however, to deposits alone. As mentioned earlier, the high absorbers in the group (particularly Indonesia, Algeria, and Iran) have made extensive use of the borrowing facilities available in the Euromarkets. Table 8-6 shows OPEC borrowings from the medium-term, syndicated Eurocurrency market during 1973 to 1976. As can be seen, OPEC borrowings decreased sharply to 3 percent of total Eurocredits in 1974 before increasing to 14 percent in 1975 and 16 percent in the first half of 1976. The pattern of OPEC borrowing is consistent with declining revenues and rising commitments in 1975–1976 and is in line with OPEC preference for borrowing now when the OPEC liquidity position is solid.

OPEC Investments in the United Kingdom

OPEC investments in the United Kingdom fell from $7.2 billion (11 percent of total surplus) in 1974 to $0.2 billion in 1975 and to net disinvestment of −$0.8 billion in the first half of 1976. The dramatic decline in OPEC investments in

Table 8-6
Eurocurrency Bank Credits to OPEC, 1972–1976
($ millions)

	1972	*1973*	*1974*	*1975*	*1976 First Half*
Total Eurocurrency credits	6857	21,851	29,263	20,992	13,724
Credits to OPEC	933	2751	1067	2900	1502
OPEC as percent of total	14	12	3	14	16
of which Algeria	172	1302	–	500	446
Indonesia	93	192	669	1348	680
Iran	335	722	115	265	231

Source: Morgan Guaranty Trust Company, *World Financial Markets* (November 1976), p. 12. Reprinted with permission.

the United Kingdom reflects the internal problems of the country as well as those relating to the steady depreciation of sterling. In recent times, OPEC bank deposits, Treasury bills, and government bonds have experienced net disinvestment, while the amounts invested in equities, property, and direct foreign currency loans have declined.

To a large extent, OPEC disenchantment with the United Kingdom derives from the large OPEC sterling holdings, whose value has been declining steadily over a period of years. In 1974 the increase in OPEC sterling holdings amounted to 2.2 billion and financed the major part of the country's current account deficit of 3.6 billion and overall balance-of-payments deficit of 3.7 billion. In 1975, however, there was a net disinvestment of OPEC sterling holdings of 180 million, while British overall deficit still amounted to some 1.7 billion.[11] OPEC alone accounts for nearly half of sterling reserves held abroad—much of it in official hands. Table 8–7 shows OPEC reserves held in sterling during 1974–1975.

The other reason behind declining OPEC investments in the United Kingdom is the declining interest rates and security yields in the country in line with movements elsewhere, especially in the United States. Table 8–8 shows United Kingdom interest rates and security yields during 1974 to 1976. As can be seen, both short-term and long-term rates declined sharply during the period, but long-term rates are more attractive than are the short-term rates. The time-yield curves of British government stocks, both medium-term and long-term, fell steadily largely because of heavy public sector borrowing, while short-term interest rates fell in response to declining trends in major money market centers of the world.[12]

OPEC Investments in the United States

OPEC investments in the United States are on the rise recently, reflecting the stability of the country as well as the existence of sophisticated money and

Table 8–7
OPEC Sterling Holdings, 1974–1975
(millions)

	Dec. 31, 1974	Dec. 31, 1975
Exchange reserves in sterling held by OPEC central monetary institutions	3101	2839
Banking and money market liabilities to other OPEC holders	344	462
Total OPEC	3445	3301
Total World	7134	7310

Source: Bank of England, *Quarterly Bulletin*, Vol. 16, No. 1 (March 1976), Tables 20/1 and 20/2.

Table 8–8

United Kingdom Short-term Interest Rates and Security Yields, 1974–1976

(percent per annum)

	Dec. 1974	Dec. 1975	Feb. 1976
Government stocks, 5 years	13.51	11.40	9.92
Government stocks, 10 years	16.58	13.31	12.22
Government stocks, 20 years	17.39	14.65	13.60
Treasury bills (3 months)	11.24	10.89	8.95
Sterling certificates of deposit (3 months)	12.69	11.17	8.75
Interbank sterling deposits (3 months)	12.56	10.72	8.78

Source: Bank of England, *Quarterly Bulletin*, Vol. 16, No. 1 (March 1976), Tables 24, 25, and 27.

capital markets. In 1974 nearly 22 percent of total OPEC surplus ($12 billion) was placed in the United States, three-quarters of which were in short-term assets, mainly Treasury bills and bank certificates of deposit. In 1975, due to smaller OPEC surplus, the United States received 16 percent (nearly $10 billion) of total investable funds, 90 percent of which took the form of longer-term investments in bonds and notes, equities, real estate, and other direct investment. In 1976, however, the share of the United States in the placement of OPEC surplus increased enormously, standing at about 44 percent of total in the first half of 1976, three-quarters of which have been channeled into longer-term investment instruments.[13]

The shift of OPEC funds in the United States from short-term to longer-term assets reflects mainly interest rate differentials. Short-term interest rates have fallen sharply in the United States in recent times, while long rates are more stable, though yielding less than in 1974. Table 8–9 shows the movement of interest rates in the United States during 1973–1976. As can be seen, the rates on Treasury bills and bank certificates of deposit have declined substantially, while yields in government and corporate bonds have held firm.

Table 8–10 shows the net short-term and long-term lending/borrowing position of banks in the United States vis-à-vis OPEC. The net short-term borrowing position of banks in 1974 amounted to more than $10 billion in 1974, but it declined in 1975 and the increase so far in 1975 has been slight. By contrast, the long-term position of United States banks vis-à-vis the Middle East has changed from net lending in 1974 to net borrowing in recent times.

It is difficult to get an idea of the precise breakdown of OPEC investments in America because of the lack of publicized data on the subject. However, a sketchy account can be provided of OPEC holdings of United States government securities. Table 8–11 shows holdings of United States government securities by a category called "Other Asia," that is, Asia excluding Japan, between January 1975 and June 1976. As can be seen, total holdings of "Other Asia"

Table 8-9
Interest Rates in the United States, 1973-1976
(percent per annum)

	Dec. 1973	Dec. 1974	Dec. 1975	Sept. 1976
Treasury bills (3 months)	7.54	7.28	5.27	5.18
Bank certificates of deposit (3 months)	9.25	9.25	5.50	5.25
Government bond yields (20 years)	7.35	8.13	8.05	7.79
Corporate bond yields (as industrial bonds with five-year call)	7.75	9.25	8.55	8.15

Source: Morgan Guaranty Trust Company, *World Financial Markets* (October 1976), pp. 15–19. Reprinted with permission.

Table 8-10
Net Borrowing (−)/Net Lending (+) Position of United States Banks vis-à-vis OPEC, 1974-1976
($ millions)

	Dec. 1974	Dec. 1975	Aug. 1976[a]
Total net short-term	−10,897	−10,530	−11,849
Venezuela	− 2709	− 2203	− 1589
Ecuador	− 122	− 120	n.a.
Indonesia	− 1064	− 264	− 894
Middle East OPEC	− 4303	− 5937	− 7266
African OPEC	− 2699	− 2008	− 2100
Net long-term (Middle East OPEC)	+ 290	− 674	− 207

Source: *Federal Reserve Bulletin* (October 1976), Tables A64–A68.

[a]Provisional.

increased from $2.9 billion in January 1975 to $8.1 billion in June 1976—much of it attributable to Middle East OPEC countries, no doubt.[14] According to the Treasury Department, during the first eight months of 1976 the OPEC nations invested $8 billion in United States stocks and bonds, amounting to 35 percent of their portfolio investments in all countries.[15] The Treasury figure probably understates the magnitude of OPEC investments, as many of them are channeled through untraceable Swiss bank trust accounts.

Table 8-12 shows net purchases of United States long-term securities, including marketable United States Treasury bonds and notes, United States corporate stocks, and United States corporate bonds, by Middle Eastern oil-exporting countries during 1975-1976. Middle East purchases of United States government securities have increased substantially during this period of time, while those of corporate stocks have held steady. Net purchases of United States corporate bonds appear to have declined a little, reflecting OPEC's preference for "risk-free" government bonds, despite the latter's lower yield.

Table 8–11

"Other Asia" Holdings of United States Government Securities, 1975–1976

($ millions, end of the month)

	Short Term	Marketable Bonds	Total
January 1975	2626	308	2934
February 1975	2698	518	3216
March 1975	1706	1043	2749
April 1975	1916	1093	3009
May 1975	2309	1268	3577
June 1975	1957	1374	3331
July 1975	2066	1374	3440
August 1975	2185	1454	3639
September 1975	2483	1605	4088
October 1975	2999	1755	4754
November 1975	3604	1806	5410
December 1975	3780	1907	5687
January 1976	3829	2136	5965
February 1976	4029	2327	6356
March 1976	3609	2859	6468
April 1976	3687	3179	6866
May 1976	4462	3640	8102
June 1976	3927	4262	8181

Source: U.S. Treasury *Bulletin*, various issues.

Table 8–12

Net Purchases of United States Long-term Securities by Middle East OPEC[a], 1975–1976

($ millions)

	1975	1976 (Jan. to Aug.)[b]
Marketable U.S. Treasury bonds and notes	1797	2659
Corporate stocks	1640	1438
Corporate bonds	1553	952

Source: *Federal Reserve Bulletin* (October 1976), Tables A68–A69.

[a]Includes Bahrain and Oman.

[b]Provisional.

Cumulative direct investment in the United States by OPEC probably amounted to more than $2 billion in 1976 out of a total of $26.7 billion in book value of foreign direct investment in United States.[16]

OPEC Development Aid and Investments

A major area where OPEC surplus funds have been utilized is in financial transfer to nonoil developing countries, including bilateral grants and loans and multilateral contributions to development-oriented international organizations, especially via purchases of World Bank and regional development bank bonds and notes and financing of the IMF Oil Facility. In 1975 OPEC development activities, as defined above, accounted for nearly one-quarter of total financial surplus, compared to 13 percent in 1974.[17]

Total bilateral and multilateral disbursed aid by OPEC increased from approximately $1 billion in 1973 to $4.6 billion in 1974 and $5.6 billion in 1975.[18] The bilaterally disbursed portion of this aid in the form of grants and loans amounted to nearly $1 billion in 1973, $3.2 billion in 1974, and $4.0 billion in 1975.[19] Total OPEC disbursed aid amounts to slightly less than 3 percent of combined GNP of OPEC members, while the concessional part of it represents slightly less than 2 percent of the combined OPEC GNP—figures much higher than the comparable ones provided by the Development Assistance Committee (DAC) of the Organization for Economic Cooperation and Development (OECD).[20] Nearly 90 percent of OPEC bilateral aid is concessional in nature, although roughly the same proportion is concentrated in Arab countries of Asia and Africa as well as in non-Arab countries of Asia with substantial Islamic populations. Table 8–13 shows the distribution of OPEC aid by donor countries in 1974, while Table 8–14 shows the geographical distribution of OPEC aid by recipients during 1973–1975.

In August 1976 OPEC established a $800 million fund to provide interest-free, long-term loans to developing countries for balance-of payments support or for special development projects, including a $400 million contribution to the International Fund for Agricultural Development. The $800 million fund

Table 8–13
OPEC Aid to Developing Countries, 1974[a]
($ millions)

Country	Concessional Aid Commitments	Disbursements	Financial Transfers Commitments	Disbursements	Total Disbursements
Algeria	117	35	22	5	40
Iran	1270	400	990	309	709
Iraq	415	235	275	–	235
Kuwait	796	384	525	246	630
Libya	377	95	482	32	127
Nigeria	12	11	240	240	251
Qatar	148	50	156	27	77
Saudi Arabia	1456	810	1534	993	1803
UAE	573	135	499	134	269
Venezuela	111	60	865	405	465
Total	5275	2215	5588	2391	4606

Source: The Society of International Development, *Survey of International Development*, Vol. 13, No. 1 (Jan.–Feb. 1976), p. 4.

[a]OPEC Members not included: Indonesia, Gabon, and Ecuador.

Table 8–14

Distribution of OPEC Bilateral Aid by Recipients, 1973–1975

($ billions)

		Percent of Total	
Region[a]	*1973*	*1974*	*First Half, 1975*
Arab Countries	93.7	77.4	79.1
of which Egypt	49.3	28.9	52.3
Jordan	8.0	5.9	3.4
Syria	15.5	17.7	9.0
Other Asia	2.9	13.7	12.4
of which India	–	7.5	–
Pakistan	2.8	6.2	10.7
Other Africa	3.4	5.0	0.8
Western Hemisphere	–	1.7	7.7
of which Argentina	–	–	2.3
Brazil	–	0.8	1.0
Guyana	–	0.9	–
Unallocated	–	0.2	–
Total	100.0	100.0	100.0
of which Concessional	90.8	92.9	62.2
Nonconcessional	9.2	7.1	37.8

Source: UNCTAD document TD/188/Supp. 1/Add. 1, Annex I, p. 27.

[a]*Arab countries*: Algeria, Bahrain, Democratic Yemen, Egypt, Jordan, Lebanon, Libyan Arab Republic, Mauritania, Morocco, Oman, Somalia, Sudan, Syrian Arab Republic, Tunisia, United Arab Emirates, Yemen. *Other Africa*: Benin, Burundi, Chad, Equatorial Guinea, Ethiopia, Gambia, Guinea, Guinea-Bissau, Madagascar, Mali, Niger, Rwanda, Senegal, Togo, Uganda, Upper Volta, Zaire, Zambia. *Other Asia*: Afghanistan, Bangladesh, Hong Kong, India, Indonesia, Pakistan, Republic of Korea, Sri Lanka. *Western Hemisphere*: Argentina, Bermuda, Brazil, Costa Rica, El Salvador, Guatemala, Gyana, Honduras, Mexico, Nicaragua,, Panama.

represents about half of the additional money required by developing countries to pay their combined annual oil bills since oil prices were increased by another 10 percent in October 1975.[21] In the IMF meeting in Manila in October 1976, OPEC members tentatively agreed to donate their share of the IMF gold sales to the IMF Trust Fund to assist 61 developing countries to overcome their balance-of-payments difficulties.[22]

In recent times OPEC countries have enlarged their assistance programs through institutions created specifically for the transfer of development funds. Chief among such institutions that exist now are the Arab Fund for Economic and Social Development, Abu Dhabi Fund for Arab Economic Development, Saudi Fund for Development, Kuwait Fund for Arab Economic Development, Arab Bank for Economic Development in Africa, Islamic Development Bank, and Special Arab Fund for Africa.[23]

OPEC investments in developing countries are mostly direct rather than portfolio—a phenomenon that is attributable to the lack of adequate capital markets in the non-Western world. Since developing countries offer attractive long-term investment opportunities in such areas as petroleum-related and other extractive industries, lately OPEC has taken an interest in investing in oil refinery and fertilizer complexes, as evidenced by Saudi involvement in South Korea and Iranian involvement in India, to cite a few examples. In addition, there have been other direct investments channeled through large international companies for reasons of safety and secrecy.[24]

OPEC multilateral development financing activities concern mostly purchases of World Bank obligations, cofinancing projects with the World Bank group, and contributions to the IMF Oil Facility. OPEC purchases of World Bank bonds and notes amounted to 13 percent of total official purchases of such World Bank debt instruments in the fiscal year 1973, 31 percent in 1974, and 57 percent in 1975 before falling sharply in 1976. In fiscal years 1974 and 1975, OPEC nations purchased $565 million and $1984 million respectively of World Bank obligations. In 1976, however, official OPEC purchases fell to $445 million.[25] Table 8–15 shows World Bank direct borrowings from OPEC countries in the fiscal year 1975–1976.

In addition, OPEC has been actively involved in cofinancing projects in developing countries, particularly with the World Bank Group. In fiscal year 1975, OPEC nations committed $60 million in cofinancing projects in the Middle

Table 8–15
World Bank Borrowings from OPEC, Fiscal Years 1975 and 1976
($ millions)

Placements	Issue	United States Dollar Amount
1975		
Total official		1984
Iran	8% loan due 1986/87	150
Nigeria	8% loan due 1988/89	240
Oman	8% loan due 1989/90	30
Saudi Arabia	8% 10-year bonds, due 1984	140.8
Saudi Arabia	8.5% 10-year bonds, due 1984	750
Trinidad & Tobago	8% loan due 1979	5
Venezuela	8% loan due 1979/89	100
Venezuela	8% loan due 1979/89	400
Other Bonds (2 years)	–	168
1976		
Total Official	–	445
Libya	7.50% 5-year bonds due 1980	38.1
Saudi Arabia	7.75% bonds due 1982/84	115.1
Saudi Arabia	8.25% bonds due 1982/86	38.4
Other Bonds (2 years)	–	253
Private		
Kuwait	8% 10-year notes	155
Total World Bank Borrowings, 1975		3501
Total World Bank Borrowings, 1976		3811

Source: World Bank, *Annual Report 1975*, p. 77, and *Annual Report 1976*, p. 84.

East and North African region.[26] The 1976 fiscal year figure is much larger; it includes an $11 million loan from the Saudi Fund for a grain storage project in Yemen, a $35 million loan from the Arab Fund for Economic and Social Development for textile manufacturing complexes in Egypt, and other loans from the Arab Bank for Economic Development in Africa, the Kuwait Fund for Arab Economic Development, and Libyan Arab Foreign Bank.[27] Table 8-16 details OPEC participation in cofinancing projects with the World Bank in fiscal year 1976. It shows OPEC interest in capital-intensive, infrastructural projects, especially in agriculture and transportation, as well as the clear geographical preference for Asian and African countries. It should also be noted that two of the cofinancing projects were in Indonesia, itself an OPEC member.

As far as the IMF Oil Facility is concerned, OPEC countries are the chief contributors, with Saudi Arabia alone financing one-third of total contributions by donor countries. OPEC loans to this facility are made in local currencies, in a basket of foreign currencies, or in SDRs; they carry an annual interest rate of 7.25 percent. Such loans are medium-term, to be repayable in 3 to 7 years.[28] Table 8-17 shows the preponderant role of OPEC contributions in the facility totaling SDR 6.9 million so far.

Table 8–16
OPEC Cofinancing Projects with World Bank, Fiscal Year 1976

Sector	Number of Projects	Beneficiaries
Agriculture	9	Burundi, Gambia, Ghana, Senegal, Somalia, Upper Volta, Yemen
Electricity	2	Indonesia, Nepal
Industry	2	Egypt, Indonesia
Technical Assistance	1	Sudan
Transportation	6	Madagascar, Mauritania, Niger, Rwanda, Yemen, Yugoslavia
Water and Sewage	2	Syria, Zaire

Source: World Bank, *Annual Report 1976*, pp. 55–72.

Table 8–17
IMF Borrowings for 1974 and 1975 Oil Facilities
(*millions of SDRs*)

| Lender | Amounts of Borrowing Agreements | | Amounts Borrowed | | Remaining Amounts May 1976 | Total |
| | For 1974 Oil Facility | For 1975 Oil Facility[a] | Fiscal years ended April 30 | | | |
			1975	1976		
Abu Dhabi	100.0	–	80.494	19.506	–	100.000
Austrian National Bank	–	100.0	–	68.650	31.350	100.000
National Bank of Belgium	–	200.0	–	100.000	100.000	200.000
Canada	246.9	–	194.301	52.627	–	246.928
Deutsche Bundesbank	–	600.0	–	455.100	144.900	600.000
Central Bank of Iran	580.0	410.0	479.760	510.240	–	990.000
Central Bank of Kuwait	400.0	285.0	333.645	313.145	38.210	685.000
Netherlands	150.0	200.0	122.000	228.000	–	350.000
Nigeria	100.0	200.0	80.815	219.185	–	300.000
Bank of Norway	–	100.0	–	64.890	35.110	100.000
Central Bank of Oman	20.0	0.5	16.286	4.214	–	20.500
Saudi Arabian Monetary Authority	1000.0	1250.0	822.600	1427.400	–	2250.000
Sveriges Riksbank	–	50.0[b]	–	20.990	29.010	50.000
Switzerland[b]	–	250.0[b]	–	191.640	58.360	250.000
Central Bank of Trinidad and Tobago	–	10.0	–	10.000	–	10.000
Central Bank of Venezuela	450.0	200.0	369.350	280.650	–	650.000
Total	3046.9[c]	3855.5	2499.251	3966.237	436.940	6902.428

Source: International Monetary Fund, *Annual Report 1976*, Appendix I, Table I.16, p. 91.

[a]Totals of agreements concluded in 1975 and 1976.

[b]The equivalents of SDR 150 million from Switzerland and SDR 100 million from the Swiss National Bank.

[c]Of which an amount equivalent to SDR 464.077 million was made available for the 1975 oil facility (from Abu Dhabi SDR 19.506 million, Canada SDR 52.627 million, Iran SDR 85.24 million, Kuwait SDR 51.355 million, the Netherlands SDR 28.0 million, Nigeria SDR 19.185 million, Oman SDR 3.714 million, Saudi Arabia SDR 123.8 million, and Venezuela SDR 80.65 million).

9 Secondary Recycling via Commercial Banks

It is not exactly accurate to say that petrodollars have been recycled smoothly without any adverse impact on the international financial system. Massive inflows of petrodollars into major commercial banks of the West, especially those dealing in the Eurocurrency markets, resulted in excess liquidity in the commercial banking sector—a problem that has been compounded by that of finding creditworthy borrowers on the assets side in the face of the global softness of business demand for bank credit throughout the world. In 1974 petrodollars saturated the Eurocurrency markets and, despite the marked slowdown of oil dollar deposits in the Euromarkets as well as in national bank certificates of deposit in 1975 and 1976, the net borrowing position of major commercial banks monitored by the Bank for International Settlements (BIS) vis-à-vis OPEC has been widening each year. Many commercial banks dealing in the Eurocurrency markets remain highly liquid and highly competitive with one another.[1] Faced with the twin problems of high liquidity and lack of creditworthy borrowers around the world, major commercial banks have resorted to questionable lending to "high-risk" customers, especially in the nonoil developing countries.

A key issue affecting the "crisis of confidence" that has engulfed the international banking community in recent times relates to the wisdom of massive foreign currency lending to less creditworthy borrowers. The future stability, or the lack of it, of international banking operations will be shaped to a large extent by how major commercial banks evaluate and manage the risks inherent in making "cross-border," foreign currency loans.

Much of the recent increase in commercial bank lending to foreigners has taken place through the Eurocurrency markets centered at London but also operating out of non-European "offshore" banking shells located in the Caribbean, Hong Kong, and Singapore. The latest statistics compiled by the BIS attest to the changing geographical pattern of Eurocurrency operations.[2] Ever since the oil price–related turmoil of 1973–1974 and largely because of slack domestic demand for bank credits in a period characterized by "stagflation," commercial bank lending to foreign nonbanks through the medium-term, internationally syndicated Eurocurrency markets has been increasing at an astonishing rate. Such lending increased from a mere $6.8 billion in 1972 to $21.8 billion in 1973 and $29.2 billion in 1974, before leveling off to still a high level of $20.9 billion in 1975. Eurolending has continued at a brisk level so far in 1976, with $12.7 billion having been committed in the first half of 1976 alone.[3]

One major change that has taken place in the pattern of publicized Eurocurrency lending is that many more loans are now being made to developing countries than to developed countries. Developing countries, both oil-producing and nonoil-producing, used to receive less than a quarter of such loans a few years ago. Now they are receiving more than half of the global total of such credits. Conversely, the share of developed countries in total loans has fallen from 71 percent in 1974 to 34 percent now. The share of communist countries has registered a marked increase from 4 percent of total loans in 1974 to 14 percent now. Table 9-1 shows the distribution of Eurocurrency medium-term credits by recipients.

The clue to understanding the radically different pattern of international private lending lies in depressed conditions of business demand for bank credits at home. It is not that bankers are suddenly kinder to the poorer countries of the world, or that the role of financing the needs of such countries is suddenly thrust upon them.[4] It is rather that banks cannot possibly sit on idle funds, which have to be put to profitable use somehow, somewhere. And a Eurocurrency loan made to a less prime borrower at at least a 9 percent rate is a great deal more profitable than investing in domestic Treasury or corporate obligations yielding much less.

Since 1974, domestic business loans made by United States commercial banks, for example, have fallen sharply, and indications are that such a decline will continue into 1977. The higher cost of bank credits relative to alternative sources of financing—internal generation of funds, commercial papers, and bond issues—has kept corporate borrowers from turning to banks.[5] Excess liquidity in the banking sector has coincided with the massive balance-of-payments financing needs of developing countries.

Table 9-1
Publicized Eurocurrency Bank Credits to Developing Countries, 1972-1976

	1972	*1973*	*1974*	*1975*	*1976 (First Half)*
Total ($ millions)	6857	21,851	29,263	20,992	12,709
Developed Countries (%)	60	63	71	34	34
Developing Countries (%)	36	33	24	53	53
OPEC (%)	14	12	3	14	16
Non-OPEC (%)	22	21	21	39	37
of which Brazil	8	3	6	10	8
Mexico	3	7	3	11	6
Peru	2	2	1	1	0.3
Philippines	1	1	3	2	6
South Korea	1	1	0.5	2	2
Others[a]	7	6	8	13	14
Socialist Countries	4	4	4	12	14

Source: Morgan Guaranty Trust Company, *World Financial Markets* (June 1976), p. 14. Reprinted with permission.

Note: Totals may not add up to 100 because of rounding.

[a]Includes regional development organizations.

Of course, developing countries come in all kinds and all shapes. About one-sixth of total credits extended to developing countries by Eurobanks has gone to oil-producing countries—chiefly, Iran, Algeria, and Venezuela—that are high absorbers of funds and that need to borrow from the Euromarkets to maintain the tempo of their grandiose development plans in the face of recession-induced declining oil-export receipts. In addition, Indonesia borrows heavily mainly to bail out Pertamina, the troubled, state-owned oil agency that accumulated debts in excess of $10 billion in the process of diversifying into nonoil ventures.[6]

The largest market for bank foreign lending, however, is nonoil developing countries, which receive nearly 40 percent of the global total of Eurocredits. Such countries have been borrowing heavily from international sources, including private sources, to finance their current account deficit, which reached a staggering $37.5 billion at the end of 1975. Between 1973 and 1975 alone they borrowed some $40 billion in medium- and long-term loans from foreign sources, half of which was provided by commercial banks.[7]

The problems of nonoil developing countries are manifold and have been discussed elsewhere.[8] The increasing involvement of commercial banks in financing current account imbalance of nonoil developing countries is controversial, to say the least.[9] In developed and developing countries alike, concern has been expressed about the sheer size of the external debt of such countries which is now in the neighborhood of $170 billion, about the relatively high proportion of commercial bank debt approaching $70 billion in total external debt, and about the massive size of the current account deficit (excluding official transfers) of such countries which is expected to be only slightly less in 1976 (about $30 billion).[10] Since many of the medium-term Eurocurrency loans already made to developing countries will mature at about the same time in early 1980s (the so-called bunching problem), calls have been made by many developing countries in UNCTAD IV that met in Nairobi in May 1976 for a general rescheduling of commercial bank debt over a period of at least 25 years. While this proposal has been rejected by lenders, at least three developing countries—Argentina, Peru, and Zaire—are known to have met with international bankers about refinancing of their existing debt.[11]

No one, however, expects an outright default by a group of developing countries on bank loans. Countries do not like to default mainly because of the impact of such an action on credit rating. Default is one thing; having difficulty in paying back is another. The irony of the developing country debt situation is that only those higher-income and middle-income developing countries that could borrow from the Eurocurrency market bacause of their creditworthiness appear to be in trouble in meeting their loan obligations. By contrast, the lower-income countries, precisely because they are poor, do not qualify for Euroloans, and many of them are not net borrowers from the commercial bank market.[12] Table 9-2 shows that Latin American countries (Brazil, Mexico, Chile, Argentina, Peru, and Colombia) are the biggest borrowers from the

Table 9-2

Estimated External Debts of Selected Non-OPEC Developing Countries, 1975

($ billions)

	External Debt		Liabilities to Commercial Banks	
	Total[a]	*Public Sector Only*[c]	*All Banks*[d]	*U.S. Banks*[e]
Brazil	21.9	10.8	14.8	9.2
Mexico	19.0	9.7	13.5	9.9
Argentina	7.3	3.5	3.2	2.1
Chile	5.2[b]	4.5	0.8	0.6
Peru	4.5[b]	3.1	2.3	1.5
Colombia	3.7[b]	2.5	1.6	1.3
South Korea	7.6	5.2	3.3	2.6
Taiwan	4.7[b]	3.2	2.1	1.8
Turkey	4.5[b]	3.4	1.0	n.a.
Philippines	3.8	1.5	2.0	1.8
Thailand	1.8[b]	0.6	1.2	0.8
Zaire	1.5[b]	1.3	0.8	n.a.
Subtotal	85	49	46.6	31.6
Other countries	65[b]	51	12.7	6.9
Total	150[b]	100	59.3	38.5

Source: Morgan Guaranty Trust Company, *World Financial Markets* (September 1976), p. 11 Reprinted with permission.

[a]Estimated outstanding disbursed external debt of public and private sectors; individual country data are not entirely comparable; some countries, indicated by (b), do not report all external debt of the private sector.

[b]See footnote a.

[c]Outstanding disbursed external debt of public sector, including private sector debt guaranteed by public sector and use of IMF credit, with original maturity of more than one year.

[d]Banks in the Group of Ten countries (which includes the United States) plus Switzerland and foreign branches of U.S. banks in Bahamas, Cayman Islands, Panama, Hong Kong, and Singapore.

[e]U.S. offices and foreign branches of U.S. banks plus U.S. offices of foreign banks.

Eurocurrency markets, accounting for two-thirds of the private debt outstanding, followed by countries in the Asia-Pacific region (South Korea, Taiwan, and Philippines).

It is no secret that private bank credits tend to be concentrated in selected developing countries—those with relatively high per capita income, close relations with the industrial world, and resource-extracting or manufacturing potential. More than two-thirds of Eurocredits made to nonoil developing countries is concentrated in a handful of developing countries—Brazil, Mexico, Argentina, Peru, Zaire, South Korea, Taiwan, and Philippines. The debt-service payment problem to private creditors, that is, amortization and interest costs, is all the more acute for precisely these relatively affluent countries in the developing world. Many of these countries continue to have double-digit debt-service ratios, that is, debt-service payments as percentage of exports. Some of them—Mexico and Peru, for example—have debt-service ratios in the neighborhood of 20 percent, a figure considered by prudent bankers to be a danger signal. Yet such countries continue to receive massive amounts of Eurocredits.[13]

As far as the United States commercial banks are concerned, it is true that long-term, nonguaranteed debt of developing countries amounts to only 5 percent of the combined assets of major banks, and that the foreign loan-loss problem is much less serious than that for domestic lending.[14] A survey by Robert Morris Associates reveals that net charge-offs of foreign loans reported by large United States banks amounted to one-fifth of domestic figures.[15] But such foreign losses are higher today than they were before, and the real magnitude of this problem is not known even to the bankers, let alone the public. The trade performance of some of the biggest bank debtors—Brazil, Mexico, Peru, Philippines, and South Korea—has been poor recently, and all are now running rather large trade account deficits.[16] As we know today, the possibility of individual developing country rescheduling of bank debt cannot be ruled out. According to a report by the United States Comptroller of Currency, substandard foreign loans by United States commercial banks probably amount to a figure in the neighborhood of $2.3 billion. A recent survey of international bankers reveals that the majority expects substantial problems on the front of the debt owed by developing countries.[17]

The strategy of United States commercial banks is diversification of risk through predetermined ceilings and geographical distribution of loans. Yet, as mentioned earlier, Euroloans to developing countries are heavily concentrated in higher-income and middle-income countries that do have a debt-service problem vis-à-vis private bankers. In the case of Citibank, for example, 20 percent of its total loan portfolio is concentrated in such countries.[18]

Despite high liquidity, the Eurocurrency market continues to a lenders' market rather than a borrowers' market. Lending terms continue to harden, reflecting greater caution about capital adequacy and creditworthiness of borrowers. Greater emphasis is being put on self-financing projects (as evidenced by the tough negotiations over the Hong Kong Mass Transit loan) and on cofinancing projects with the World Bank and other public development agencies than on freewheeling program lending based on government guarantees alone. Many banks are relying more and more on the performance criteria of IMF's standby arrangements and are experimenting with novel lending approaches such as syndicate-to-syndicate loans.[19]

Shorter maturities and higher interest rates are here to stay. Less creditworthy borrowers in the developing world pay a multitiered interest rate that includes a floating margin over the base rate or the London Interbank offered rate. In addition, banks charge a variety of front-end fees, including participation fees for internationally syndicated loans. The effective cost of borrowing, therefore, is much higher for less creditworthy borrowers than for prime borrowers. As in the past, the poor are paying a high price for the problems of the rich as well as for their borrowing needs.

10 OPEC Dependence on the Western World

From the standpoint of oil-consuming countries, the strains on OPEC unity that appeared at the Doha meeting in December 1976 are less significant in the short run than the fact that oil prices have been raised again by the cartel. Despite ostensible breaches in the solidarity of OPEC, the basic bias of the cartel is toward raising prices, as the majority of the cartel members are high-absorbing countries that need additional revenues to keep up with massive development spending needs. The low-absorbing members of the group—the countries in the Arabian Peninsula led by Saudi-Arabia—are an exception to this rule, mainly because they do not need additional revenues and are concerned about the long-term impact of higher oil prices on economic and political stability in the West, which will shape both the future demand for OPEC oil and the value of OPEC investments which have been building up in the West for some time. But even the Saudis are sensitive to the development aspirations of the cartel partners and, in view of their strong feelings about the political stalemate in the Middle East, may find it wise not to break away from the cartel.

Thus, while the eventual future of the cohesiveness of the oil cartel is a moot question, in the short run the oil-consuming world can expect a series of relatively small oil price increases almost every year. The ultimate aim of the high-absorbing members of OPEC is probably to stabilize oil prices at around $20 per barrel by 1980.[1] The leverage of the cartel derives more from the increasing Western dependence on OPEC oil than from any basic solidarity of the group as such, which is necessarily composed of a diverse conglomeration of developing countries with differing views, capacities, and needs.[2] In the absence of a substantial reduction in Western consumption demand for OPEC oil or a sizable breakthrough in developing non-OPEC sources of energy—both unlikely phenomena in the immediate future on economic and technological grounds—Western leverage on inevitable oil price increases is limited. Polemics about "taking on" OPEC appear futile without reducing domestic demand for imported oil or increasing domestic supply of energy.[3]

One might be tempted to assume, therefore, that the West will be getting the wrong end of the stick in its future dealings with OPEC. The financial realities behind OPEC oil power, however, point to precisely the opposite kind of conclusion. While there is no denying the fact that the economic impact of higher oil prices on the Western world is substantial by way of diversion of income from current consumption, reduced consumer demand for goods and services, shaken investors' confidence, deteriorating growth rates of GNP, increasing balance-of-payments difficulties, and related structural changes in the economy,[4] much of the adverse economic impact on selected Western countries, led by the United States, Japan, and West Germany, can be and has been cushioned to a large extent by higher Western exports to OPEC as well as by increasing OPEC investments in the West. No amount of petrodollar recycling, however, can solve the basic problems of inherently weaker Western economies with large and persistent balance-of-payments deficits, noncompetitive exports, inferior investment opportunities, or other structural inadequacies, many of which precede the quintupling of oil prices since late 1973. OPEC assets in Britain, for example, have registered net disinvestment in recent times, reflecting the internal problems of the country and the steady depreciation of the pound sterling.[5]

It might be legitimately argued, though, that the recycling process does indeed involve certain economic and political costs that might be unacceptable to the West. Some Western countries might be reluctant to give up more exports in order to obtain a given amount of OPEC oil. Others might be concerned about foreigners buying up productive capacities of Western societies. Still others might worry about the increasing gap between the rich and the poor countries within the Western world itself—a process that is being accentuated by the concentration of OPEC financial surplus in low-risk areas of the world.[6] While these concerns are understandable, the fact is that the West has little choice in this matter, as long as it continues to rely on OPEC oil for its energy needs.

While the "real" aspects of higher oil prices portend high costs to Western oil-consuming countries, the financial aspects of the recycling of petrodollars are manageable and have carried few of the doomsday dangers that had been predicted earlier.[7] The manageability of the recycling issue is due both to the fundamental dependence of the OPEC countries on the Western world in trade and investment areas and to the relative neutrality and pragmatism of the international financial markets in reshuffling funds back and forth between oil producers and oil consumers of the world. Going one step further, one can even say that the financial aspects of petrodollars imply a "blessing in disguise" for the more advanced Western countries. The United States, Japan, and West Germany dominate the rather specialized import market of OPEC, and the United States, a country noted for massive infusions of foreign capital throughout its past history, is gaining again from increasing OPEC investments in longer-term, capital-intensive sectors.

As far as import-export recycling is concerned, OPEC dependence on a handful of Western suppliers for its import needs means a booming export business for particularly the United States, Japan, and West Germany. The industrial countries as a group provide 80 percent of OPEC's imports, nearly 65 percent of which is accounted for by six countries (the United States, Japan, West Germany, France, the United Kingdom, and Italy), and nearly 45 percent of which is supplied by the three Western giants—the United States (18 percent),

Japan (15 percent), and West Germany (12 percent).[8] To the extent that many OPEC imports are either resource-intensive goods (food and foodstuffs) or capital-intensive products (computers, electronics, aircrafts, and equipment relating to agriculture, communication, construction, oil field and mining, vehicle and repair, and so on), the number of potential suppliers of such highly specialized goods is necessarily limited, even within the Western world. The higher the degree of technology required for the production of such products, the less competitive is the market for exporters of such goods. The possibility of a Western "cartel" on exports to OPEC, involving such symptoms as oligopolistic control and price fixing, therefore cannot be ruled out. Indeed, in certain export items such a phenomenon has already occurred. Moreover, the changing technology required for many of these specialized products leaves OPEC few options but to continue relying on a limited number of Western suppliers for the needed inputs for finished goods. Clearly, over the long run, the shift of capital from the West to OPEC will be minimized, and possibly reversed, to the extent that selected Western countries can gain a significant economic advantage through their specialized exports.

As far as OPEC exports are concerned, industrial countries presently absorb 80 percent of OPEC oil exports and slightly less than half of OPEC's nonoil exports such as rubber, tin, cocoa, copper, carpets, soap, clothing, knitwear, shoes, hides and leather, agricultural raw materials, and so on. Western markets also are vital for OPEC's diversification strategies, because they will absorb in the future the vast proportion of OPEC exports of natural gas, petrochemicals, iron and steel, and "invisibles" such as shipping and oil tankers.[9] Without Western markets there is little future for such specialized OPEC products.

The dependence of OPEC nations on the Western world is even greater in the area of the investment of their surplus oil revenues. For the low absorbers of the group, chiefly the countries in the Arabian Peninsula, there is necessarily a limit to the domestic investment possibilities of excess revenues. Four OPEC countries (Saudi Arabia, Kuwait, United Arab Emirates, and Qatar) account for 90 percent of OPEC financial surplus and possess nearly half of OPEC's foreign exchange reserves. Yet, domestic spending in such countries amounts to only about one-third of their oil-export earnings. These are precisely the countries that can cut their present oil production level in half without jeopardizing their revenue needs.[10] Almost by definition, therefore, petrodollars accumulated by such countries must be recycled back to the West through OPEC investments in domestic financial markets of the West as well as through the Eurocurrency markets. However, as mentioned earlier, not all Western countries will benefit equally from OPEC investments abroad, depending upon the attractiveness, or the lack of it, of money and capital markets that exist in individual Western countries. By the same token, the non-Western, nonoil developing countries will probably be bypassed in this recycling process due to inferior investment opportunities, defined as the absence of active secondary markets, that exist there.

The nature of OPEC investments in the West has changed dramatically since the latter part of 1975 because of profound changes in international interest rates. In 1974 the vast portion of OPEC's investable surplus was recycled through international short-term money market instruments, mainly via bank certificates of deposit and Treasury bills, because they coincided with abnormally high interest rates on short-term deposits that prevailed in that year. The high short-term rates of 1974 were a reflection of excess demand for short-term credits and tight official monetary policies that existed in a number of Western countries— developments that had preceded the influx of oil money. In 1974 OPEC investors had the best of all worlds: high return on short-term deposits, sufficient demand for their surplus funds from quality borrowers in the West, and limited risks. To the extent that there were risks, they were being assumed by banks, which in bidding for petrodollar deposits, also incurred the liability for their profitable and creditworthy placements. Short-term investments fitted in nicely with OPEC's preference for high liquidity and great flexibility in maturities. There was no reason for OPEC to explore longer-term investment opportunities in view of the interest rate differential alone. In July 1974, for example, a 5-year United States Treasury note was yielding approximately 8½ percent, or 400 basis points below the 90-day commercial bank CD rate or 5/8 percent below the 3-month Treasury bill rate. Also, investing petrofunds in longer-term ventures such as corporate equity participation would have taken months, if not years, to work out, especially in view of the prevailing resistance in the West to the absorption of the newly found wealth in the hands of a group of non-Western countries. It is not surprising, therefore, that as much as 66 percent of total OPEC financial surplus in 1974 was kept on a short-term basis in the Eurocurrency markets, the United States, and the United Kingdom. Such investments were made in predominantly sterling or dollar-denominated instruments, because these were the only currencies that had sufficient circulating volume to accommodate the massive injection of petrofunds. By the same token, the international money markets came to be highly dependent on such short-term OPEC investments for their liquidity, causing great concern in many governments of the Western world. It was feared by some observers that such massive short-term deposits could be withdrawn on call or week-fix basis, causing an immediate financial, and perhaps political, chaos throughout the Western world.[11]

The main reason why such pessimistic forecasts have not come true is that, while OPEC nations control the price of oil, they do not control international interest rates, which are determined by demand and supply of capital as well as by official intervention policies. Since the worldwide recession of 1974-1975, interest rates in major money and capital markets of the world have skidded largely because of the abrupt decline in demand for credit from low-risk, quality corporate customers. The precipitous fall in interest rates paid for credit means a lower rate of return paid to investors with capital. Short-term investment opportunities are no longer the bargain that they used to be for OPEC just a couple of years ago. Yields in short-term United States Treasury obligations have dropped dramatically, and so have those on Eurocurrency deposits. Moreover, the pound sterling, which represented a large portion of OPEC foreign currency reserves and was a major vehicle for OPEC's foreign investments in 1974, has depreciated in value by 20 percent during the past few years. Even gold prices have fallen sharply from the peak attained in 1974. Such changes in international rates of return on investment possibilities are beyond the control of even the oil-rich OPEC investors.

In July 1974, 3-month Treasury bills were yielding approximately 9 1/8 percent; by October 1976, that rate declined to slightly above 5 percent. In July 1974, 90-day certificates of deposit at commercial banks in the United States were yielding as high as 12½ percent or a spread of 3 3/8 percent above what the Treasury was paying on a comparable investment. By October 1976, bank deposits were paying only 5¼ percent, or 25 basis points above the 90-day Treasury bill rate. Although all investment alternatives are yielding much less than in 1974, the Treasury yield curve is now more positively sloped and offers a higher rate of return for longer-term investments. Thus, OPEC investors are being obliged to commit their funds for longer periods of time and for lower rates of return than were available two years ago. In other words, OPEC investors in the United States and elsewhere are getting "locked in" by having to give up short-term control over their funds as well as running the liquidity and market risks involved in making longer-term investments.

The evolution in the pattern of OPEC investments in the United States is indeed striking. In 1974, nearly 75 percent of OPEC investments in the United States (totaling $12 billion, or 22 percent of total OPEC financial surplus) was kept in short-term assets, mainly in Treasury bills and commercial bank certificates of deposit. By the first half of 1976, nearly 75 percent of OPEC's investable surplus placed in the United States (44 percent of total surplus) had poured into longer-term investment instruments such as bonds and notes, equities, real estate, and other forms of direct investment.[12] In January 1975, out of a total of nearly $3 billion invested by OPEC in government securities, $2.6 billion was held in Treasury bills with maturities under 1 year. By August 1976, OPEC held $4.2 billion in Treasury bills and $4.8 billion in Treasury notes and bonds with maturities of beyond 1 year, for a total investment of $9 billion. Many OPEC holdings of government notes and bonds are now committed for an average life of 2 to 5 years, with the majority of such investments being held in the 18 months to 3 years range.

Whether they like it or not, the commitment of surplus petrodollars in longer-term investment vehicles has forced the high absorbers of OPEC to be more dependent on short-term borrowings from the international banking system to meet their immediate liquidity needs either in paying for imports or in meeting budgetary shortfalls. As a group, OPEC borrowings of internationally syndicated, medium-term Eurocurrency credits increased from a mere 3 percent of the global total of such credits extended in 1974 to 16 percent in the first half of 1976, with the lion's share being absorbed by three OPEC countries—Iran, Indonesia, and Algeria.[13] As for the low absorbers in the cartel, longer-term investments in the West are making them more vulnerable to economic disruptions in the world which might affect the value of their cumulative investments abroad. Indeed, since OPEC's long-term aspirations are so intimately tied to economic and political stability in the West, the day might very well come (and indications are that it did indeed come at the Doha meeting last December) when some members of OPEC will realize that, for the sake of their investments alone, they have more to gain by maintaining a reasonably stable price of oil than by arbitrarily raising it at regular intervals. After all, one does not kill the goose that lays the golden egg.

The policy implications of the OPEC dependence on the Western world are clear. By the 1980s, OPEC nations are expected to import nearly $150 billion worth of goods and services from abroad. The United States ranks first as the biggest supplier of goods and services into OPEC countries and dominates the three largest OPEC markets, namely, those of Iran, Saudi Arabia, and Venezuela. American exports to OPEC are expected to increase substantially in the future, although perhaps at a slower rate of growth than has been the case so far. At the same time, OPEC investments in the United States have risen substantially in recent times, amounting to slightly less than half of the cartel's financial surplus. Such foreign investments are beneficial to America in that they provide relatively "painless" funds for the Treasury's massive borrowing needs as well as the financing needs of the corporate sector.

A policy of confrontation with OPEC hardly makes sense for a country that is a real winner in the recycling game and that is becoming more and more dependent on the OPEC component of imported oil. OPEC investments in the West are consistent with the economic principle of the marginal efficiency of capital and are meant to be channeled into financial assets or productive capacities to ensure the survival of oil-producing countries in the postoil era. By the same token, the exploration of petrodollars as a source of long-term capital for corporate needs in America is not yet fully realized. There is no reason why the trickle of petrofunds—less than what America spends each year on liquor or cosmetics—cannot be absorbed fruitfully in the trillion-dollar economy of the United States.

Appendix

Table A-1
World Crude Oil Production, 1973-1975

	1973		1974		1975[a]	
	1000 barrel/day	Percent	1000 barrel/day	Percent	1000 barrel/day	Percent
Total	56,180	100.0	55,855	100.0	53,140	100.0
North America	11,452	20.4	11,145	20.0	10,500	19.9
United States	9189	16.4	8870	15.9	8870	15.8
Canada	1798	3.2	1695	3.0	1470	2.8
Mexico	465	0.8	580	1.0	720	1.4
Central and South America	5131	9.1	4220	7.6	3600	6.8
Venezuela[b]	3364	6.0	2970	5.3	2350	4.4
Ecuador[b]	204	0.4	160	0.3	170	4.4
Other	1563	2.8	1090	2.0	1080	2.0
Western Europe	370	0.7	370	0.7	550	1.0
United Kingdom	Negl.	Negl.	Negl.	Negl.	20	Negl.
Norway	30	0.1	30	0.1	190	0.4
Other	340	0.6	340	0.6	340	0.6
Communist Countries	9895	17.6	10,720	19.2	11,650	21.9
U.S.S.R.	8420	15.0	9020	16.1	9630	18.1
Other	1475	2.6	1700	3.0	2020	3.8
Africa	5902	10.5	5330	9.5	4960	9.3
Algeria[b]	1070	1.9	940	1.7	920	1.7
Libya[b]	2187	3.9	1520	2.7	1520	2.9
Nigeria[b]	2053	3.7	2260	4.0	1780	3.3
Gabon[b]	150	0.3	180	0.3	210	0.4
Other	442	0.8	430	0.8	530	1.0
Asia-Pacific	2272	4.0	2320	4.1	2240	4.2
Indonesia[b]	1339	2.4	1380	2.5	1310	2.5
Other	933	1.7	940	1.7	930	1.8
Middle East	21,158	37.7	21,750	38.9	19,580	36.8
Saudi Arabia[b,c]	7607	13.5	8480	15.2	7080	13.3
Kuwait[b,c]	3024	5.4	2550	4.6	2080	3.9
Iran[b]	5861	10.4	6040	10.8	5350	10.1
Iraq	1964	3.5	1820	3.3	2250	4.2
UAE	1298	2.3	1410	2.5	1400	2.6
Qatar[b]	570	1.0	520	0.9	430	0.8
Other	834	1.5	930	1.7	990	1.9

Source: *International Economic Report of the President*, Transmitted to the Congress in March 1976 (Washington: Government Office, 1976), p. 172.

[a]Estimate.
[b]OPEC member.
[c]Includes one-half of Neutral Zone Production.

Table A-2
United States Oil and Gas Self-sufficiency and Reserve Ratio, 1955-1974

Year	Petroleum (thousand barrels/day)				Natural Gas (billion cubic feet)				Reserve to Production Ratio	
	Consumption[a]	Production	Imports	Self-sufficiency (%)	Consumption[a]	Production	Imports[b]	Self-sufficiency (%)	Crude[c]	Natural Gas
1955	8826	7578	1248	85.9	9405	9405	–	100.0	12.4	22.1
1956	9413	7977	1436	84.7	10,064	10,064	–	100.0	11.9	21.8
1957	9552	7978	1574	83.5	10,718	10,680	38	99.6	11.8	21.4
1958	9217	7517	1700	81.6	11,166	11,030	136	98.8	12.9	22.1
1959	9712	7932	1780	81.7	12,180	12,046	134	98.9	12.8	21.1
1960	9806	7991	1815	81.4	12,927	12,771	156	98.2	12.8	20.1
1961	10,091	8174	1917	81.0	13,473	13,254	219	98.4	12.6	19.9
1962	10,435	8353	2082	80.0	14,278	13,877	401	97.2	12.3	20.0
1963	10,763	8640	2123	80.3	15,153	14,747	406	97.3	11.9	19.0
1964	11,058	8800	2258	79.5	15,904	15,462	442	97.2	11.7	18.3
1965	11,482	9014	2468	78.5	16,497	16,040	457	97.2	11.7	17.6
1966	12,152	9579	2573	78.8	17,687	17,207	480	97.3	11.3	16.5
1967	12,757	10,220	2537	80.1	18,735	18,171	564	97.0	10.3	15.9
1968	13,476	10,636	2840	78.9	19,974	19,322	652	96.7	9.8	14.8
1969	13,993	10,827	3166	77.4	21,425	20,698	727	96.6	9.3	13.3
1970	14,716	11,297	3419	76.8	22,741	21,920	821	96.4	8.9	12.1
1971	15,081	11,155	3926	74.0	23,428	22,493	935	96.0	8.7	11.5
1972	15,970	11,229	4741	70.2	23,551	22,532	1019	95.7	8.1	10.7
1973	17,202	10,946	6256	63.6	23,681	22,648	1033	95.6	8.1	9.9
1974	16,541	10,453	6088	63.2	22,674	21,715	959	95.8	8.1	9.9

Source: Robert E. Higgins, "The Oilfield Services Industry," Goldman Sachs *Investment Research* (June 22, 1976), p. 6. Reprinted with permission.

[a]Including natural gas liquids.
[b]Almost solely from Canada.
[c]Excluding 9.6 billion barrels located in Prudhoe Bay, Alaska. Reserve life including such reserves is as follows: 1971, 11.7; 1972, 11.7; 1973, 11.1; 1973, 11.1; 1974, 11.3.

Table A-3
United States Crude Oil and Natural Gas Production: Total and Offshore, 1955-1974

Year	Crude Oil Production (thousand B/D)				Natural Gas Production (billion cubic feet)			
	Total U.S.[a]	Total U.S. Offshore		Increment Offshore as a Percent of Increment Total	Total U.S.	Total U.S. Offshore		Increment Offshore as a Percent of Increment Total
		Amount	Percent of Total U.S.			Amount[a]	Percent of Total U.S.	
1955	6807	162	2.4		9405	128	1.4	
1956	7151	201	2.8	11.3	10,064	143	1.4	2.3
1957	7170	229	3.2	147.4	10,680	174	1.6	5.0
1958	6710	236	3.5	b	11,030	258	2.3	24.0
1959	7054	274	3.9	11.0	12,046	353	2.9	9.4
1960	7035	320	4.5	b	12,771	440	3.4	12.0
1961	7183	365	5.1	30.4	13,254	478	3.6	7.9
1962	7332	444	6.1	53.0	13,877	640	4.6	26.0
1963	7542	515	6.8	33.8	14,747	763	5.2	14.1
1964	7664	589	7.7	60.7	15,462	850	5.5	12.2
1965	7804	665	8.5	54.3	16,040	939	5.9	15.4
1966	8295	823	9.9	32.2	17,207	1373	8.0	37.2
1967	8810	1009	11.5	36.1	18,171	1838	10.1	48.2
1968	9096	1291	14.2	98.6	19,322	2321	12.0	42.0
1969	9237	1441	15.6	106.4	20,698	2845	13.7	38.1
1970	9637	1577	16.4	34.0	21,920	3218	14.7	30.5
1971	9463	1685	17.8	b	22,493	3751	16.7	93.0
1972	9441	1684	17.8	c	22,532	3757	16.7	15.4
1973	9209	1596	17.3	c	22,648	3975	17.6	187.9
1974	8765	1460	16.7	c	21,715	4230	19.5	b

[a]Includes lease condensate.
[b]Total production declined while offshore increased.
[c]Total production and offshore production both declined.

Source: Robert E. Higgins, "The Oilfield Services Industry," Goldman Sachs *Investment Research* (June 22, 1976), p. 16. Reprinted with permission.

Table A-4
Estimated Expenditures for Exploration, Development, and Production of Oil and Gas in the United States, 1959-1974 (*$ millions*)

	1959	1960	1961	1962	1963	1964	1965	1966	1967	1968	1969	1970	1971	1972	1973	1974
Exploration																
Drilling and equipping exploratory wells[a]																
Dry-hole costs[a]	821	774	774	847	790	854	849	832	802	826	944	815	775	910	1021	1647
Acquiring undeveloped acreage	554	626	428	815	376	570	438	577	829	1578	1137	714	642	1722	3646	5659
Lease rentals and expense for carrying leases	193	193	189	197	193	177	166	180	140	179	134	138	143	142	155	186
Geological and geophysical	320	277	280	299	300	336	355	378	392	373	387	349	361	372	429	640
Contributions toward test wells											33	30	24	35	38	34
Land department, leasing, and scouting	124	104	115	108	117	100	102	70	82	82	93	98	100	105	102	117
Other, including direct overhead		71	65	58	69	72	61	128	122	136	168	143	142	147	181	231
Subtotal	2012	2045	1851	2324	1845	2109	1971	2165	2371	3174	2896	2287	2187	3433	5572	8514
Development																
Drilling and equipping development wells[a]																
Drilling and equipping producing wells[a]	1830	1651	1624	1729	1512	1574	1553	1528	1497	1583	1634	1733	1573	1869	2016	2686
Lease equipment	483	431	446	537	527	619	580	878	991	968	442	443	388	497	524	770
Improved recovery programs											303	285	323	310	276	399
Other, including direct overhead											180	170	185	160	189	349
Subtotal	2313	2082	2070	2266	2039	2193	2133	2406	2488	2551	2559	2631	2469	2836	3005	4204
Production																
Production costs, including direct overhead	1450	1390	1455	1535	1581	1613	1685	1895	1933	2094	2189	2379	2504	2563	2792	3508
Production or severance taxes	316	339	346	354	373	393	400	430	464	499	525	563	587	613	683	129
Ad Valorem taxes	192	199	195	202	198	204	212	212	248	259	271	294	295	269	275	384
Subtotal	1958	1928	1996	2091	2152	2210	2297	2537	2645	2852	2985	3236	3386	3445	3750	4021
Overhead																
Exploration	183	197	219	213	200	215	207	195	206	204	210	189	206	239	293	387
Development	442	424	457	478	470	461	487	478	495	539	207	220	202	257	250	272
Production											369	416	465	467	485	584
Subtotal	625	621	676	691	670	676	694	673	701	743	786	825	873	963	1028	1243
Total expenditures	6908	6676	6593	7372	6706	7188	7095	7781	8205	9320	9226	8979	8915	10,677	13,355	17,982

Source: Robert E. Higgins, "The Oilfield Services Industry," Goldman Sachs *Investment Research* (June 22, 1976), p. 21. Reprinted with permission.

[a]Joint Association Survey questionnaire was amended as of 1969, reclassifying certain costs between exploration and development.

Table A-5
Non-OPEC Developing Countries: Crude Oil Production
(*thousands of barrels/day*)

	1975			1977[a]			1980[a]		
	Onshore[b]	Offshore[b]	Total	Onshore	Offshore	Total	Onshore	Offshore	Total
Higher-income									
Argentina	395.0	–	395.0	400.0	–	400.0	395.0	–	395.0
Bahrain[d]	61.0	–	61.0	60.0	–	60.0	55.0	–	55.0
Brazil	170.0	21.0	191.0	130.0	100.0	230.0	110.0	230.0	360.0
Brunei[d]	35.0	140.0	175.0	30.0	200.0	250.0	25.0	215.0	260.0
Chile[c]	37.0	–	37.0	35.0	15.0	50.0	32.0	18.0	50.0
China, Rep. of	5.0	–	5.0	4.0	–	4.0	4.0	–	4.0
Columbia	164.0	–	164.0	150.0	–	150.0	135.0	–	135.0
Israel	n.a.	n.a.	75.0	n.a.	n.a.	3.0	n.a.	n.a.	3.0
Malaysia[d]	–	90.0	90.0	–	200.0	200.0	n.a.	320.0	320.0
Mexico[c,d]	785.0	65.0	806.0	930.0	70.0	1000.0	1125.0	75.0	1200.0
Oman[d]	342.0	–	342.0	360.0	–	360.0	305.0	–	305.0
Peru	35.0	40.0	75.0	125.0	45.0	170.0	165.0	35.0	200.0
Trinidad & Tobago	48.0	157.0	205.0	41.0	154.0	195.0	33.0	167.0	200.0
Tunisia	75.0	20.0	95.0	70.0	30.0	100.0	60.0	60.0	120.0
Subtotal	2152.0	328.0	2720.0	2335.0	834.0	3172.0	2444.0	1160.0	3607.0
Middle-Income									
Angola[d]	23.0	142.0	166.0	25.0	150.0	175.0	25.0	200.0	255.0
Bolivia[d]	42.0	–	42.0	50.0	–	50.0	60.0	–	60.0
Congo[d]	–	38.0	38.0	–	35.0	35.0	–	45.0	45.0
Egypt[d]	55.0	175.0	230.0	135.0	315.0	450.0	115.0	415.0	530.0
Syria[d]	170.0	–	175.0	250.0	–	250.0	300.0	–	300.0
Subtotal	290.0	355.0	651.0	460.0	500.0	960.0	500.0	660.0	1160.0
Lower-Income									
Burma	20.0	–	20.0	30.0	–	30.0	35.0	–	35.0
India	165.0	–	165.0	250.0	20.0	225.0	195.0	200.0	395.0
Pakistan	6.0	–	6.0	7.0	–	7.0	7.0	–	7.0
Zaire[d]	–	2.0	2.0	–	50.0	50.0	–	50.0	50.0
Subtotal	191.0	2.0	193.0	242.0	70.0	312.0	237.0	250.0	487.0
Total	2533.0	685.0	3560.0	3037.0	1404.0	4444.0	3181.0	2070.0	5254.0

Source: Adrian Lambertini, "Energy Problems of the Non-OPEC Developing Countries, 1974–80," *Finance and Development*, Vol. 13, No. 3 (September 1976), p. 27.

[a]Projected.
[b]Estimated Breakdown.
[c]Includes condensate.
[d]Net oil-exporting country.

Table A-6
Crude Oil Production in Iran, 1913–1975
(*thousand barrels*)

Year	Daily Average	Total	Cumulative	Annual % Change in Daily Production
1913	5.0	1825	1825	
1914	8.0	2920	4745	60.0
1915	10.0	3650	8395	25.0
1916	12.0	4392	12,787	20.0
1917	19.0	6935	19,722	58.3
1918	23.6	8623	28,345	24.2
1919	27.8	10,139	38,484	17.8
1920	33.4	12,230	50,714	20.1
1921	45.7	16,637	67,387	36.8
1922	61.0	22,247	89,634	33.5
1923	69.1	25,230	114,864	13.3
1924	88.5	32,373	147,237	28.1
1925	96.0	35,038	182,275	8.5
1926	98.2	35,842	218,117	2.3
1927	108.7	39,688	257,805	10.7
1928	118.7	43,461	301,266	9.2
1929	115.5	42,145	343,411	− 2.7
1930	125.6	45,833	389,244	8.7
1931	121.6	44,376	433,620	− 3.2
1932	135.2	49,471	483,091	11.2
1933	149.0	54,392	537,483	10.2
1934	158.5	57,851	595,334	6.4
1935	156.9	57,273	652,607	− 1.0
1936	171.4	62,718	715,325	9.2
1937	213.2	77,804	793,129	24.4
1938	214.7	78,372	871,501	0.7
1939	214.1	78,151	949,652	− 0.3
1940	181.2	66,317	1,015,969	−15.4
1941	139.1	50,777	1,066,746	−23.2
1942	198.0	72,256	1,139,002	42.3
1943	204.4	74,612	1,213,614	3.2
1944	278.8	102,045	1,315,659	36.4
1945	357.6	103,526	1,446,185	28.3
1946	402.2	146,819	1,593,004	12.5
1947	424.7	154,998	1,748,002	5.6
1948	520.2	190,384	1,938,386	22.5
1949	560.9	204,712	2,143,098	7.8
1950	664.3	242,475	2,385,573	18.4
1951	349.6	127,600	2,513,173	−47.4
1952	27.6	10,100	2,523,273	−92.1
1953	26.8	9800	2,533,073	− 2.9

85

Table A-6 (Cont.)

Year	Daily Average	Total	Cumulative	Annual % Change in Daily Production
1954	61.4	22,400	2,555,473	129.1
1955	328.9	120,035	2,675,508	435.7
1956	541.8	198,289	2,873,797	64.7
1957	719.8	262,742	3,136,539	32.9
1958	826.1	301,526	3,438,065	14.8
1959	928.2	338,810	3,776,875	12.4
1960	1067.2	390,766	4,167,641	15.0
1961	1202.2	438,804	4,606,445	12.6
1962	1334.5	487,084	5,093,529	11.0
1963	1491.3	544,325	5,637,854	11.7
1964	1710.7	626,107	6,263,961	14.7
1965	1908.3	696,520	6,960,481	11.6
1966	2131.8	778,109	7,738,590	11.7
1967	2603.2	950,180	8,688,770	22.1
1968	2839.8	1,039,367	9,728,137	9.1
1969	3375.8	1,232,155	10,960,292	18.9
1970	3829.0	1,397,585	12,357,877	13.4
1971	4539.5	1,656,918	14,014,795	18.6
1972	5023.1	1,838,455	15,853,250	10.7
1973	5860.9	2,139,229	17,992,479	16.7
1974	6021.6	2,197,901	20,190,380	2.7
1975	5350.1	1,952,787	22,143,167	−11.2

Source: Courtesy information provided by the National Iranian Oil Company.

Table A-7

Geographical Distribution of Exports of Crude Oil and Petroleum Products by the Trading Companies Affiliated with the Oil Service Company of Iran
(*Percent*)

	Crude Oil					Petroleum Products				
	1971	1972	1973	1974	1975[a]	1971	1972	1973	1974	1975
Total	100.0	100.0	100.0	100.0	100.0	100.0	100.0	100.0	100.0	100.0
Western Europe	27.2	35.3	41.2	44.5	46.6	5.9	4.2	4.9	14.0	15.3
Japan	46.4	41.7	34.9	26.9	27.1	22.5	15.8	14.8	14.4	14.0
Asia	8.9	7.5	7.2	5.0	2.3	16.7	17.9	19.8	21.8	22.8
Central and North America	9.1	9.5	11.7	16.7	15.0	2.9	3.2	3.7	3.0	3.2
Africa	7.0	4.3	3.5	5.3	6.8	28.4	24.2	19.8	12.3	10.3
Australasia	0.4	0.3	0.2	0.2	0.7	7.9	9.5	8.6	6.8	5.7
South America	0.6	0.2	0.3	0.5	0.4	2.0	0	1.2	2.0	2.7
Other regions	0.4	1.2	1.0	0.9	1.1	13.7	25.2	27.2	25.7	26.0

Source: Bank Markazi Iran, *Annual Report and Balance Sheet 2534*, p. 92.

[a]Preliminary estimate.

Notes

Notes

Introduction

1. Essentially, the OPEC viewpoint on oil is that it is a valuable raw material which should not be burned as mere fuel but should be used to produce value-added products, a vast variety of which could form petrochemical intermediaries. Also, OPEC maintains that higher oil prices are a "blessing in disguise" in that they have at last focused the attention of the international community on developing alternative energy fuels rather than relying on plentiful supplies of still relative relatively "cheap" but depletable oil. Finally, OPEC affirms that higher oil prices are a means of fostering a more "rational" division of labor and more "equitable" terms of trade in the world economy. See the address by Gumersindo Rodriguez, Governor of the Bank for Venezuela, at the Joint Session of the World Bank-IMF Annual Meeting in 1975 in IMF, *Summary Proceedings, Annual Meeting, 1975*, p. 8; address by Abdlatif Y. Al-Hamad, Director-General of the Kuwait Fund for Arab Economic Development, at the Bankers Trust Company Anniversary Conference in London in September 1974, reproduced in *International Development Review*, Vol. 17, No. 3 (1975/3), pp. 10-13; and comments by Francisco Parra, former Secretary-General of OPEC, in *The Oil Daily* (October 14, 1975), pp. 1-2.

2. *Recycling of Petrodollars*, Hearings before the Permanent Subcommittee on Investigations of the Committee on Government Operations, U.S. Senate, 93d Congress, 2d Session, October 16, 1974 (Washington: U.S. Government Printing Office, 1974), pp. 78-79; "OPEC's Import Costs and Crude Oil Price Increases," Petroleum Industry Research Foundation, New York (October 26, 1976), pp. 1-10; Morgan Guaranty Trust Company, *World Financial Markets* (September 1976), p. 5.

3. Address by Abdlatif Y. Al-Hamad (see note 1); and Statement of Felix R. Dias Bandaraniake, Governor of the Fund and Bank for Sri Lanka, in IMF *Summary Proceedings, Annual Meeting 1975*, p. 219.

4. Address by Abdlatif Y. Al-Hamad (see note 1); and Address by Gumersindo Rodriguez (see note 1).

5. Statement by Manuel Perez-Guerrero, former Secretary-General of UNCTAD, to the Committee of Twenty, 18 January 1974, Rome, reproduced in UNCTAD, *Monthly Bulletin*, No. 91 (March 1974). For an analysis of OPEC-Third World solidarity, see Branislav Gosovic and John Gerard Ruggie, "On the Creation of a New International Economic Order: Issue Linkage and the Seventh Special Session of the UN General Assembly," *International Organization*, Vol. 30, No. 2 (Spring 1976), pp. 309-45.

6. First National City Bank, "Why OPEC's Rocket Will Lose Its Thrust," *Monthly Economic Letter* (June 1975), p. 11.

7. For an indication of the "alarmist" views expressed by Western leaders, see Dankwart A. Rustow and John F. Mugno, *OPEC: Success and Prospects* (New York: New York University Press, 1976), pp. 51-58.

Chapter 1
Western Dependence on OPEC OIL: The Basis of Recycling

1. General Agreement on Tariffs and Trade, *Press Release*, GATT 1183 (27 August 1976), p. 15.

2. *IMF Survey* (March 15, 1976), p. 91; and Morgan Guaranty Trust Company, *World Financial Markets* (September 1976), p. 5.

3. *International Economic Report of the President*, Transmitted to the Congress in March 1976 (Washington: Government Printing Office, 1976), pp. 108-109.

4. Ibid., pp. 13, 171.

5. Morgan Guaranty Trust Company, *World Financial Markets* (September 1975), p. 5.

6. *International Economic Report of the President*, pp. 16, 109. The United Kingdom North Sea oil reserves are estimated at about 9 billion barrels and are capable of producing between 1 and 1.5 million barrels per day by 1980. Norway can produce another 700,000 barrels per day by 1978. See *Mobil World*, Vol. 40, No. 3 (March/April 1974), p. 7. For details on Trans-Alaskan pipeline, see Goldman Sachs, *Investment Research* (July 29, 1976), pp. 13, 13a.

7. *International Economic Report of the President*, pp. 13, 179-180. Both Venezuela and Canada have reduced oil shipments to the United States to meet their domestic priorities. Venezuela's current oil production has actually declined, and there are political as well as environmental objections in the country to tapping potential oil reserves. Canada is a net importer of petroleum, and there is also political resistance in Canada to serving insatiable United States oil needs. Nigeria and Indonesia produce insignificant amounts of crude oil (3.3 and 2.5 percent of global production respectively) relative to Western needs, and Indonesian oil production declined in 1975 by 4.2 percent over the 1974 figure. In any case, their markets are not primarily geared to the United States. For details see Robert E. Hunter, *The Energy Crisis and U.S. Foreign Policy*, Overseas Development Council development paper 14 (August 1973), pp. 22-32; and Robert E. Higgins, "The Oilfield Services Industry," Goldman Sachs *Investment Research* (June 22, 1976), p. 46.

8. Speech by His Excellency Shaikh Ali A. Alireza, Ambassador of Saudi Arabia to the United States, at National Foreign Trade Convention, New York, November 16, 1976.

9. The dependence of the United States on net imports as a percentage of domestic consumption has been increasing steadily over the years, rising from 23 percent in 1970 to 36 percent in 1973, 37 percent in 1974, 38 percent in 1975, and 40 percent in 1976. Over the past 10 years or so, domestic oil consumption in the United States increased 51 percent, but domestic production increased by only 20 percent. Although the rate of growth in demand for oil has slowed slightly in recent times, there is still a substantial gap between domestic production and consumption. There has been a corresponding decline in United States reserves of oil to production ever since 1959. For details see: *World Oil Developments and U.S. Oil Import Policies*, Senate Committee on Finance, 93d Congress, 1st Session, December 12, 1973 (Washington: Government Printing Office, 1973), pp. 8-14; Remarks by Herman J. Schmidt, Vice Chairman, Mobil Oil Corporation, at 28th Annual Convention of the National Business Aircraft Association, New Orleans, October 29, 1975, and at Los Angeles Society of Financial Analysts, June 22, 1976. Remarks by A. E. Murray, President, U.S. Marketing and Refining Division, Mobil Oil Corp, at the 74th Annual Meeting of the American Automobile Association, San Francisco, September 29, 1976; Testimony of William P. Tavoulareas, President of Mobil Oil Corporation, before the Energy Subcommittee of the Joint Economic Committee of the U.S. Congress, June 2, 1976.

10. The United States has had oil import control ever since 1955. The Mandatory Oil Import Program, based on quotas, remained in effect from 1959 to April 1973, when it was replaced by the present system of license fees, which still uses the former quotas as the basis for fee-free allotments. With a view to reducing domestic oil consumption by 1 million barrels per day, President Ford imposed an import fee of $1 per barrel in January 1975 and raised it to $2 in June 1975. See *World Oil Developments and U.S. Oil Import Policies*, pp. 3-4; and "Oil and Energy," *Department of State Gist*, Bureau of Public Affairs (June 1975).

11. Remarks by Herman J. Schmidt in New Orleans.

12. Wilbur L. Gay, "Oil Industry Commentary: An Analytical Service," Goldman Sachs *Investment Research* (July 1976), pp. 17-18.

13. Robert E. Higgins, "The Oilfield Services Industry," Goldman Sachs *Investment Research* (June 22, 1976), p. 10.

14. See John Hagel, *Alternative Energy Strategies: Constraints and Opportunities* (New York: Praeger, 1976).

15. Goldman Sachs *Investment Research* (April 8, 1976), pp. 15, 15a. According to the Organization of Arab Petroleum Exporting Countries (OAPEC), the share of oil in world energy production will probably decline to 50 percent in 1990 from 55 percent in 1974, but oil consumption is expected to increase at an annual average rate of at least 4 percent. The result is that world oil production would have to increase by nearly 50 percent in 1990 over present production level. See *IMF Survey* (November 15, 1976), p. 352.

Chapter 2
OPEC Financial Surplus: The First Prong of Recycling

1. See Morgan Guaranty Trust Company, *World Financial Markets* (March 1976), pp. 10-11; and First National City Bank, "Why Opec's Rocket Will Lose Its Thrust," *Monthly Economic Letter* (June 1975), p. 12.

2. International Monetary Fund, *Annual Report 1976* (Washington: IMF, 1976), p. 13.

3. Morgan Guaranty Trust Company, *World Financial Markets* (January 21, 1976), p. 7; (September 1976), p. 6; and GATT *Press Release*, GATT 1183 (27 August 1976), p. 6.

4. Morgan Guaranty Trust, January 21, 1976; p. 8; GATT *Press Release*, p. 8; *IMF Survey* (March 15, 1976), pp. 81, 91.

5. Morgan Guaranty Trust, September 1976, pp. 4-8.

6. Morgan Guaranty Trust, March 1976, pp. 10-11.

7. W. J. Levy, *Future OPEC Accumulation of Oil Money: A New Look at a Critical Problem*, reprinted in *Financial Support Fund*: Hearings before the Committee on Foreign Relations, U.S. Senate, 94th Congress, First Session (Washington: Government Printing Office, 1976), pp. 20-45.

8. U. S. Senate Committee on Finance, *World Oil Developments and U.S. Oil Import Policies*, pp. 22-25. The CIF cost of oil includes FOB tax-paid cost plus tanker cost. Between January 1975 and June 1976, the FOB cost of oil increased by $1.05 per barrel, while CIF cost rose by $1.12 per barrel, the difference being the slight increase in tanker cost. See Goldman Sachs *Investment Research* (July 1976), p. 21.

9. U.S. Senate Committee on Finance, ibid., p. 25.

10. Morgan Guaranty Trust, September 1976, p. 5.

11. Ibid., p. 7, and January 21, 1976, p. 8; IMF, *Annual Report 1976*, p. 19; and *International Economic Report of the President*, Transmitted to the Congress in March 1976 (Washington: Government Printing Office, 1976), p. 13.

12. *IMF Survey* (February 2, 1976), pp. 43-44.

13. Morgan Guaranty Trust, September 1976, pp. 7-8; and First National City Bank, *Monthly Economic Letter* (June 1975), p. 14.

Chapter 3
Western Exports to OPEC: The Second Prong of Recycling

1. *Federal Reserve Bulletin* (April 1976), pp. 291-92.

2. Ibid.; and *Commerce America* (May 10, 1976), p. 3.

3. The development plans of OPEC countries comprise capital outlays of $142 billion for Saudi Arabia during 1975-1979; $123 billion for Iran during 1973-1978; $52 billion for Venezuela during 1976-1980; $15.2 billion for Kuwait during 1976-1980; $30 billion for Nigeria during 1975-1980; and $25 billion for Indonesia during 1974-1979. The problems of Nigeria and Indonesia are not what to do with oil money but rather the lack of enough of it in view of large population bases and small current oil output. See *Commerce America* (February 2, 1976), pp. 28-45; *Commerce America* (August 2, 1976), pp. 28-45; *Commerce America* (July 19, 1976), p. 8; *1976/77 Budget Speech* by His Excellency Lieutenant-General Olusegun Obasanjo, Head of the Federal Military Government of Nigeria, 31 March 1976; *A Study of the Relationships between the Government and the Petroleum Industry in Selected Foreign Countries: Indonesia*, Prepared by the Congressional Research Service for the Committee on Interior and Insular Affairs, U.S. Senate (Washington: Government Printing Office, 1975); and *Summary of Saudi Arabian Five-Year Development Plan (1975-1980)*, U.S. Treasury Department, reprinted in *Technology Transfer to the Organization of Petroleum Exporting Countries*, Hearings before the Subcommittee on Domestic and International Scientific Planning and Analysis of the Committee on Science and Technology, U.S. House of Representative, 94th Congress, 1st Session, October 1975 (Washington: Government Printing Office, 1976), pp. 227-313.

4. *Mitsubishi Bank Review*, Vol. 7, No. 2 (February 1976), p. 295.

5. *Federal Reserve Bulletin* (April 1976), pp. 291-92; and *Commerce America* (May 10, 1976), pp. 2-3.

6. Department of State *News Release* (May 14, 1976), pp. 2-4; *Commerce America* (August 2, 1976), pp. 31-33.

7. *Commerce America* (July 19, 1976), pp. 2-3.

Chapter 4
The Iranian Position on Oil Prices

1. In 1975 Iran produced 10.6 percent of world oil, 26.7 percent of Middle Eastern oil, and 19.5 percent of OPEC oil. See *Iran Oil Journal*, No. 189 (Spring, 1976), p. 4. Iran alone accounts for 45 percent of total OPEC arms imports. See *International Economic Report of the President*, Transmitted to the Congress in March 1976 (Washington: GPO, 1976), p. 13.

2. Bank Markazi Iran, *Annual Report and Balance Sheet 1352* (as of March 20, 1975), pp. 7-8; Press Conference of the Shah of Iran as reprinted in *Iran Oil Journal*, No. 181 (Spring 1974), pp. 3-4; and *Iran Oil Journal*, No. 184 (November 1974), pp. 3-5.

3. Speech by Dr. Manouchehr Eghbal, Chairman of NIOC, reprinted in *Iran Oil Journal*, No. 186 (Spring 1975), pp. 15-17; and Press Conference of the Shah of Iran, pp. 3-6.

4. *Iran Oil Journal*, No. 178 (July 1973), p. 3. It took Iran 22 years to move from nationalization of the Anglo-Iranian Oil Company in 1951 to gaining complete control over the oil industry in 1973-1974. For a history of the Iranian oil industry, see Jahangir Amuzegar, *Energy Policies of the World: Iran* (Delaware: Center for the Study of Marine Policy, 1975), pp. 8-44.

5. *Petroleum Industry in Iran* (Tehran: The Iranian Petroleum Institute, 1975), pp. 13-26.

6. Ibid.

7. Ibid., p. 22.

8. The 1950s and the 1960s were characterized by perpetual disputes between Iran and international oil companies regarding posted price of oil, government's share of oil revenues, and so on. See Amuzegar, *Energy Policies of the World*, pp. 32-33, 44.

9. *Ibid.*, pp. 31-32.

10. *Ibid.*, p. 38.

11. *Iran Oil Journal*, No. 185 (February 1975), p. 5; and "Iran's Fifth Development Plan: A Short Review," *Iran Economic News* (March 1975), p. 2.

Chapter 5
The Pattern of Government Revenues and Expenditures

1. Plan and Budget Organization, The Imperial Government of Iran, *The Budget 1353: A Summary*, March 21, 1974-March 20, 1975, Part I, p. 2.

2. Bank Markazi Iran, *Annual Report and Balance Sheet, 1351* (as of March 20, 1973), pp. 5, 14, 15, 33, 46-48, and 78.

3. Ibid., p. 15.

4. Ibid., pp. 47, 168-69, 202.

5. *Petroleum Industry in Iran* (Tehran: The Iranian Petroleum Institute, 1975), p. 107; and *Iran Economic News* (March 1975), "Iran's Fifth Development Plan," pp. 1-5.

6. *Iran Economic News*, ibid., p. 6. Of the total disbursements of $123 billion, 41 percent is earmarked for current expenditures, 34 percent for fixed development investments, 9 percent for investments abroad, and 16 percent for repayment of foreign loans and other expenditures. Of the fixed investments, 66 percent would be made by the public sector and 34 percent by the private sector. See Bank Markazi Iran, *Annual Report and Balance Sheet 1353*, pp. 31-32.

7. Bank Markazi Iran, ibid., p. 33.

Chapter 6
Changing Oil Revenue and Shifting Investment Priorities

1. *The Budget 1353*, Plan and Budget Organization, The Imperial Government of Iran, pp. 1-6; and Bank Markazi Iran, *Annual Report 1352*, pp. 14-18, (as of March 20, 1974).

2. Bank Markazi Iran, *Annual Report and Balance Sheet 1353* (as of March 20, 1975), p. 11.

3. Bank Markazi Iran, *Annual Report and Balance Sheet 2534* (as of March 20, 1976), p. 2.

4. *Iran Economic News*, Vol. 2, No. 2 (February 1976), p. 3.

5. *Iran Economic News*, Vol. 1, No. 6 (June 1975), p. 4; and *Quarterly Economic Review: Iran*, No. 2 (1975), p. 8.

6. Bank Markazi Iran, *Annual Report 2534*, p. 2.

7. Ibid., pp. 12-14.

8. Interviews at Iranian Embassy, Washington, D.C.

9. First National City Bank, *Monthly Economic Letter* (June 1975), p. 14.

10. International Finance Division, Economic Analysis and Projections Department, World Bank, May 1976; and Morgan Guaranty Trust, *World Financial Markets* (October 1976), p. 12.

11. *International Financial Statistics*, Vol. 28, No. 8 (August 1975), pp. 192-93; *Quarterly Economic Review: Iran*, No. 4 (1974), p. 8; and *Iran Economic News*, Vol. 2, No. 2 (February 1976), p. 3. Iran's aid program is impressive. Disbursed aid as a percentage of Iranian GNP amounts to nearly 2 percent, although the chief beneficiaries of Iranian aid are India, Pakistan, Egypt, Turkey, Sudan, and Syria. See *Quarterly Economic Review: Iran*, No. 2 (1975), p. 9.

12. Courtesy information provided by the New York Office of NIOC; and *Iran Economic News*, Vol. 2, No. 2 (February 1976), p. 1, Vol. 2, No. 6 (June 1976), p. 2, and Vol. 2, No. 9 (September 1976), p. 1.

Chapter 7
Implications of Iran's Diversification Programs

1. Inflationary pressures had been building up in Iran ever since 1973-1974 because of the rapid rise in effective demand brought about largely by sharp increases in government expenditures and private sector liquidity. Aggregate supply of goods and services, on the other hand, could not keep up with the phenomenal rise in aggregate demand due to shortages of raw materials and infrastructural bottlenecks in the distribution of imported goods. The result was high inflation averaging between 15 and 20 percent per annum. The government

embarked on a crash program to limit aggregate demand and expand aggregate supply at the same time. Current government expenditures were reduced, domestic monetary and credit policies were regulated, prices were supervised, and the supply of imported goods was accelerated. The results of such stern measures were impressive. The average annual increase in the consumer price index was less than 10 percent in 1975-1976 compared with more than 15 percent in the previous year. The wholesale price index rose by only 4 percent in 1975-1976 compared with 16 percent in the previous year. See Bank Markazi Iran, *Annual Report and Balance Sheet 2534* (as of March 20, 1976), pp. 2-5.

2. *Iran Economic News* (Embassy of Iran, Washington), Vol. 1, No. 6 (June 1975), p. 4; Vol. 2, No. 4 (April 1976), p. 2; Vol. 2, No. 5 (May 1976), p. 3; and Vol. 2, No. 9 (September 1976), p. 4.

3. *Petroleum Industry in Iran, op. cit.*, p. 78.

4. *Iran's Fifth Development Plan* (March 1975), *op. cit.*, p. 6; and Amuzegar, *op. cit.*, pp. 47-51.

5. *Petroleum Industry in Iran* (Tehran: The Iranian Petroleum Institute, 1975), pp. 67-75; and *Iran Oil Journal*, No. 189 (Spring 1976), p. 25.

6. *Petroleum Industry in Iran*, ibid., pp. 77-91; *Iran Oil Journal*, ibid., p. 25, and No. 186 (Spring 1975), p. 13; and *Iran Economic News*, Vol. 2, No. 8 (August 1976), p. 4, and Vol. 2, No. 10 (October 1976), p. 3.

7. *Iran Economic News*, Vol. 2, No. 3 (March 1976), p. 1, and Vol. 2, No. 9 (September 1976), p. 5.

8. U.S. Department of Commerce, *Overseas Business Reports* (March 1975), p. 4; and Bank Markazi Iran, *Annual Report and Balance Sheet 2534*, p. 63.

Chapter 8
OPEC Investments Abroad

1. Morgan Guaranty Trust Company, *World Financial Markets* (September 1976), p. 7; and (March 1976), p. 11.

2. Ibid.

3. Bank for International Settlements, *45th Annual Report*, 1 April 1974-31 March 1975 (Basel, June 1975), pp. 130, 139; and *46th Annual Report*, April 1975-31 March 1976 (Basel, June 1976), pp. 77, 84.

4. BIS, *46th Annual Report*, ibid., pp. 84, 89.

5. Bank of England, *Quarterly Bulletin*, Vol. 15, No. 1 (March 1975), Table 23; and Vol. 16, No. 1 (March 1976), p. 22.

6. Hikmat Sh Nashashibi, "Surplus OPEC Funds Switch into the Euromarkets," *Euromoney* (October 1975), pp. 54-55.

7. M. M. Abushadi, "Arab Surpluses Move into Eurocredits and Bonds," *Euromoney* (November 1975), pp. 18-21.

8. Ibid.; BIS, *45th Annual Report*, p. 146; BIS, *46th Annual Report*, p. 84.

9. See A. K. Bhattacharya, *The Asian Dollar Market: International Offshore Financing* (New York: Praeger, 1977), chapters 4, 6.

10. World Bank, *Annual Report 1975*, p. 79; and *Annual Report 1976*, p. 87.

11. Bank of England, *Quarterly Bulletin*, Vol. 16, No. 1 (March 1976), p. 12.

12. Ibid., pp. 18-20, 23.

13. Morgan Guaranty Trust Company, *World Financial Markets* (September 1976), p. 7; and *Federal Reserve Bulletin* (April 1976), p. 292.

14. According to the Federal Reserve, "Other Asia" holdings of marketable United States Treasury bonds and notes increased from $212 million in 1974 to $2099 million in 1975, and to $4885 million in August 1976. Such holdings represented 40 percent of total foreign holdings in August 1976, compared with 27 percent in 1975 and 4 percent in 1974. See *Federal Reserve Bulletin* (October 1976), Table A66. Excluding government securities, OPEC's other long-term investments amount to only about 4 percent of total foreign portfolio investment in the United States. See *IMF Survey* (March 15, 1976), pp. 94-95.

15. Address by John D. Wilson, Senior Vice President, The Chase Manhattan Bank, at the 63d National Foreign Trade Convention, New York City, November 16, 1976.

16. *Department of State Gist*, Bureau of Public Affairs, May 1975, and September 1976.

17. Morgan Guaranty Trust Company, *World Financial Markets* (September 1976), p. 6.

18. World Bank, *Annual Report 1975*, p. 7, and *Annual Report 1976*, p. 10.

19. UNCTAD document TD/188/Supp.1/Add.1, Annex I, p. 25; and Morgan Guaranty Trust Company, World Financial Markets (September 1976), p. 6.

20. UNCTAD, ibid., p. 26; and Society for International Development, *Survey of International Development*, Vol. 13, No. 1 (Jan.-Feb. 1976), p. 4.

21. *IMF Survey* (August 16, 1976), p. 254.

22. *IMF Survey* (November 1, 1976), p. 336.

23. For a description of such institutions, see Society for International Development, *Survey of International Development*, p. 5; and *Business International Money Report* (May 22, 1975), p. 107.

24. A. Kapoor, "The New Middle East, the Developing Countries, and the International Company," in A. Kapoor (ed.), *Asian Business and Environment* (Princeton: The Darwin Press, 1976), pp. 581-87.

25. World Bank, *Annual Report 1975*, pp. 76–77; and *Annual Report 1976*, pp. 83–84.

26. World Bank, *Annual Report 1975*, p. 42. Although there have not been cofinancing projects in Latin America, Venezuela has made financial contributions to a special oil facility in Central America and Panama, the financing of coffee stocks in Central America, bilateral loans to Latin American countries, a trust fund in the Inter-American Development Bank, loans to the Central American Bank for Economic Integration, the Caribbean Development Bank, and the United Nations Emergency Operation. Ibid., pp. 48–49.

27. World Bank, *Annual Report 1976*, pp. 45–47.

28. IMF, *Annual Report 1975*, pp. 55–56, and *Annual Report 1976*, p. 54.

Chapter 9
Secondary Recycling via Commercial Banks

1. See A. K. Bhattacharya, "The Asian Dollar Has to Fight a Risky Reputation," *Euromoney* (September 1976), pp. 84–85.

2. Bank for International Settlements, *Forty-sixth Annual Report*, 1 April 1975–31 March 1976 (Basel, June 1976), pp. 75–76, 85–87.

3. Morgan Guaranty Trust Company, *World Financial Markets* (June 1976), p. 14.

4. For the bankers' point of view, see Citibank, "LDC Default—Shadow without Substance," *Monthly Economic Letter* (November 1976), pp. 13–15; and Morgan Guaranty Trust Company, *World Financial Markets* (May 1976), pp. 8–9.

5. Federal Reserve Bank of St. Louis, *U.S. Financial Data* (November 5, 1976), p. 11.

6. See A. K. Bhattacharya, *The Asian Dollar Market: International Offshore Financing* (New York: Praeger, 1977), chapter 4.

7. IMF *Annual Report 1976*, p. 15; Citibank, "LDC Deficits—Why the Tide Rose, Why It's Ebbing," *Monthly Economic Letter* (June 1976), p. 7.

8. See A. K. Bhattacharya, *Foreign Trade and International Development* (Lexington, Mass.: Lexington Books, D. C. Heath, 1976).

9. See A. K. Bhattacharya, "How Far Does Asia Benefit from the Asian Dollar Market?" *The Banker* (November 1976), pp. 25–27.

10. IMF *Annual Report 1976*, p. 15; World Bank, *Annual Report 1976*, pp. 14–17; Citibank (June 1976), p. 7. About half of the $50 billion debt to private banks carries a maturity of more than 1 year, and the rest is taken up by short-term debts and debt-service payments. See Morgan Guaranty Trust Company, *World Financial Markets* (September 1976), pp. 10–11.

11. UNCTAD *Monthly Bulletin*, No. 118 (June–July 1976); Remarks of Richard D. Hill, Chairman of the First National Bank of Boston, before the 63d National Foreign Trade Convention, New York City, November 15, 1976.

12. See Bhattacharya, *Foreign Trade and International Development*, pp. 24–28; and BIS, *46th Annual Report*, pp. 85–88.

13. According to the latest World Bank report, nine countries owed more than $1 billion to banks in 1974: Zaire, South Korea, Argentina, Brazil, Mexico, Peru, Algeria, Greece, and Spain. It is interesting to note that all, except Greece and Spain, have double-digit debt-service ratios: 11.7 for Zaire, 10.5 for South Korea, 16.2 for Argentina, 15.2 for Brazil, 18.4 for Mexico, 25.6 for Peru, and 14.4 for Algeria. See World Bank, *Annual Report 1976*, pp. 102–105.

14. The nonguaranteed, long-term debt of developing countries amounts to about $12 billion. See Citibank (November 1976).

15. *IMF Survey* (August 16, 1976), p. 245.

16. Morgan Guaranty Trust, *World Financial Markets* (September 1976), pp. 9–10.

17. "Banking in Asia '76," *Far Eastern Economic Review* (April 23, 1976), p. 43.

18. Citicorp., *1975 Annual Report*, p. 18. Ninety-six percent of Citicorp's exposure in developing countries is concentrated in oil-exporting and higher-income and middle-income countries. Out of a total net loan loss of $299.4 million in 1975, overseas loan losses of Citicorp amounted to $100.5 million, or about one-third of total. Ibid.

19. Remarks by Richard D. Hill.

Chapter 10
OPEC Dependence on the Western World

1. See the address by K. D. Malaviya, Indian Minister of Petroleum, in *India News*, Vol. 15, No. 34 (November 19, 1976), p. 1.

2. For an indication of the diversity of OPEC in the areas of population, per capita GNP, oil revenues, nonoil exports, and so on, see Jane W. Jacqz (ed.), *Iran: Past, Present, and Future* (New York: The Aspen Institute for Humanistic Studies, 1976), p. 316.

3. The reason why the so-called Adelman thesis—the view that the "energy crisis" is a "fiction" manipulated by OPEC as well as by international oil companies—runs into problems is that, although crude oil is now in plentiful supply, it is not realistic to expect OPEC nations to let their single most precious natural resource be depleted, especially when some 20 percent of the Western World's present demands are simply surplus to real needs. See M. A. Adelman, "Is the Oil

Shortage Real?" *Foreign Policy*, No. 9 (Winter 1972-1973), pp. 69-107; "Letters on Oil," *Foreign Policy*, No. 11 (Summer 1973), pp. 126-33; Comments by Francisco Parra, former Secretary-General of OPEC, in *The Oil Daily* (October 14, 1975), pp. 1-2; and Department of State, *Newsletter* (June 3, 1976).

Also, see the testimony of Julius L. Katz, Deputy Assistant Secretary of State for Economic and Business Affairs, in *Inventory of Economic Relations between the United States and OPEC Countries*, Hearings before the Committee on Government Operations, U.S. Senate, 94th Congress, 1st Session, July 25, 1975 (Washington: Government Printing Office, 1975), p. 24.

4. See the statement of Henry C. Wallich, Member, Board of Governors of the Federal Reserve System, in *Recycling of Petrodollars*, Hearings before the Permanent Subcommittee on Investigations of the Committee on Government Operations, U.S. Senate, 93d Congress, 2d Session, October 16, 1974 (Washington: Government Printing Office, 1974), pp. 27-32.

5. Morgan Guaranty Trust Company, *World Financial Markets* (September 1976), p. 7.

6. See the statement of William E. Simon, Secretary of the Treasury, in *Effect of Petrodollars on Financial Markets*, Hearings before the Subcommittee on Financial Markets of the Committee on Finance, U.S. Senate, 94th Congress, 1st Session, January 30, 1975 (Washington: Government Printing Office, 1975), pp. 5-8.

7. See Khodadad Farmanfarmaian et al., "How Can the World Afford OPEC Oil?" *Foreign Affairs* (January 1975), reprinted in A. Kapoor (ed.), *Asian Business and Environment in Transition* (Princeton: Darwin Press, 1976), pp. 559-80.

8. Morgan Guaranty Trust Company, *World Financial Markets* (November 1976), p. 3.

9. Bank Markazi Iran, *Annual Report and Balance Sheet 2534* (as of March 20, 1976), pp. 56-59, 91-92.

10. *IMF Survey* (February 2, 1976), pp. 43-44, and (March 15, 1976), p. 96.

11. See *Effect of Petrodollars on Financial Markets*, Hearings before the Subcommittee on Financial Markets of the Senate Committee on Finance, 94th Congress, 1st Session, January 30, 1975 (Washington: Government Printing Office, 1975).

12. Morgan Guaranty Trust Company, *World Financial Markets* (September 1976), p. 7.

13. Morgan Guaranty Trust Company, *World Financial Markets* (November 1976), p. 12.

Bibliography

Official Documents

Bank of England. *Quarterly Bulletin*. London: Bank of England, Vol. 15, No. 1 (March 1975), and Vol. 16, No. 1 (March 1976).

Bank for International Settlements. *Annual Report*. Basel: Bank for International Settlements. 45th Annual Report (June 1975), and 46th Annual Report (June 1976).

Bank Markazi Iran. *Annual Report and Balance Sheet*. Tehran: Bank Markazi Iran. 1351 (1972-1973), 1352 (1973-1974), 1353 (1974-1975), and 2534 (1975-1976).

Council on International Economic Policy, The White House. *International Economic Report of the President*, Transmitted to the Congress in March 1976. Washington: Government Printing Office, 1976.

General Agreement on Tariffs and Trade. *International Trade 1974/75*. Geneva: GATT, 1975.

_____ . "Prospects for International Trade." *Press Release*. GATT 1183 (27 August 1976).

International Monetary Fund. *Annual Report 1976*. Washington: IMF, 1976.

_____ . *IMF Survey*. Various issues.

_____ . *International Financial Statistics*. Various issues.

_____ . *Summary Proceedings, Annual Meeting 1975*.

_____ . *Finance and Development*. Various issues.

Iran Economic News. Embassy of the Imperial Government of Iran, Washington, D.C. Various issues.

Iran Oil Journal. National Iranian Oil Company, Tehran. Various issues.

Petroleum Industry in Iran. National Iranian Oil Company, Tehran, 1975.

Plan and Budget Organization. The Imperial Government of Iran. *The Budget 1354 and Amended 1353 (1974-75/1975-76): A Summary*. Tehran: PBO, 1976.

United Nations Conference on Trade and Development. *International Financial Cooperation for Development*. Geneva: UNCTAD, May 1976. TD/188, Supp. 1, and Add. 1.

_____ . *Monthly Bulletin*. Various issues.

United States Commerce Department. *Commerce America*. Various issues.

_____ . *Overseas Business Reports*. Various issues.

_____ . *Survey of Current Business*. Vol. 56, No. 8 (August 1976).

United States Congress. Senate. *Financial Support Fund*. Hearings before the Committee on Foreign Relations. 94th Congress, 1st Session, July 1975 and March 1976. Washington: Government Printing Office, 1976.

United States Congress. House. *Technology Transfer to the Organization of Petroleum Exporting Countries.* Hearings before the Subcommittee on Domestic and International Scientific Planning and Analysis of the Committee on Science and Technology. 94th Congress, 1st Session, October 1975. Washington: Government Printing Office, 1976.

_____ . House. *The Energy Crisis and Proposed Solutions.* Panel Discussions before the Committee on Ways and Means. 94th Congress, 1st Session, March 1975. Washington: Government Printing Office, 1975.

_____ . Senate. *Inventory of Economic Relations between the United States and OPEC Countries.* Hearings before the Committee on Government Operations. 94th Congress, 1st Session, July 25, 1975. Washington: Government Printing Office, 1975.

_____ . Senate. *Effect of Petrodollars on Financial Markets.* Hearings before the Subcommittee on Financial Markets of the Committee on Finance. 94th Congress, 1st Session, January 30, 1975. Washington: Government Printing Office, 1975.

_____ . Senate. *Recycling of Petrodollars.* Hearings before the Permanent Subcommittee on Investigations of the Committee on Government Operations. 93d Congress, 2d Session, October 16, 1974. Washington: Government Printing Office, 1974.

_____ . Senate. *World Oil Developments and U.S. Oil Import Policies.* A Report prepared for the Committee on Finance by the U.S. Tariff Commission. 93d Congress, 1st Session, December 12, 1973. Washington: Government Printing Office, 1973.

United States Federal Reserve. *Bulletin.* April 1976, and October 1976.

United States State Department. *Current Policy.* Bureau of Public Affairs, Office of Media Services. Various issues.

_____ . *Gist.* Bureau of Public Affairs. Various issues.

_____ . *News Release.* Bureau of Public Affairs, Office of Media Services. Various issues.

_____ . *Special Report.* Bureau of Public Affairs, Office of Media Services. Various issues.

United States Treasury Department. *Bulletin.* Various issues.

World Bank. *Annual Report 1976.*

Corporate Publications

Citibank. *Monthly Economic Letter.* Various issues.

Citicorp. *Annual Report 1975.*

Goldman Sachs. *Investment Research.* Various issues.

Mitsubishi Bank Review. Various issues.

Mobil World. Various issues.

Morgan Guaranty Trust Company. *World Financial Markets*. Various issues.

The Oil Daily. Various issues.

Petroleum Industry Research Foundation. "Oil Import Costs and Crude Oil Price Increases." October 26, 1976.

Books

Al-Otaiba, Mana Saeed. *OPEC and the Petroleum Industry*. New York: John Wiley & Sons, 1975.

Bhattacharya, Anindya K. *Foreign Trade and International Development*. Lexington, Mass.: Lexington Books, D. C. Heath, 1976.

_____. *The Asian Dollar Market: International Offshore Financing*. New York: Praeger, 1977.

Field, Michael. *A Hundred Million Dollars a Day: Inside the World of Middle East Money*. New York: Praeger, 1976.

Fried, Edward R., and Schultze, Charles L. (eds.). *Higher Oil Prices and the World Economy*. Washington: The Brookings Institution, 1975.

Hagel, John. *Alternative Energy Strategies: Constraints and Opportunities*. New York: Praeger, 1976.

Jacqz, Jane W. (ed.). *Iran: Past, Present, and Future*. New York: Aspen Institute for Humanistic Studies, 1976.

Kapoor, A. (ed.). *Asian Business and Environment in Transition*. Princeton: Darwin Press, 1976.

Mikdashi, Zuhayr. *The Community of Oil Exporting Countries: A Study in Governmental Cooperation*. Ithaca, N.Y.: Cornell University Press, 1972.

Odell, Peter R. *Oil and World Power: Background to the Oil Crisis*. Middlesex: Penguin, 1975.

Rustow, Dankwart A., and Mugno, John F. *OPEC: Success and Prospects*. New York: New York University Press, 1976.

Booklets and Articles

Abushadi, M. M. "Arab Surpluses Move into Eurocredits and Bonds," *Euromoney* (November 1975), pp. 18-21.

Adelman, M. A. "Is the Oil Shortage Real?" *Foreign Policy*, No. 9 (Winter 1972-1973), pp. 69-107.

Amuzegar, Jahangir. *Energy Policies of the World: Iran*. Delaware: Center for the Study of Marine Policy, 1975.

Amuzegar, Jahangir. "North-South Dialogue: From Conflict to Compromise," *Foreign Affairs*, Vol. 54 (April 1976), pp. 547-62.

Bhattacharya, Anindya K. "The Asian Dollar Has to Fight a Risky Reputation," *Euromoney* (September 1976), pp. 84-85.

――― . "How Far Does Asia Benefit from the Asian Dollar Market?" *The Banker*, Vol. 126 (November 1976), pp. 1225-28.

Enders, T. O. "OPEC and the Industrial Countries: The Next 10 Years," *Foreign Affairs*, Vol. 53 (July 1975), pp. 625-37.

Evans, Michael K. "The Energy Crisis and the U.S. Economy," *Euromoney* (January 1974), p. 24.

Farmanfarmaian, Khodadad, et al., "How Can the World Afford OPEC Oil?" *Foreign Affairs*, Vol. 53 (January 1975), pp. 201-22.

Gosovic, Branislav, and Ruggie, John Gerard. "On the Creation of a New International Economic Order: Issue Linkage and the Seventh Special Session of the UN General Assembly," *International Organization*, Vol. 30 (Spring 1976), pp. 309-45.

Guth, Wilfried. "International Money and the Oil Crisis," *The Banker*, Vol. 124 (November 1974), pp. 1447-49.

Hansen, Roger D. "The Political Economy of North-South Relations: How Much Change," *International Organization*, Vol. 29 (Autumn 1975), pp. 921-47.

Harrison, Stephen. "Eurodollar Banks and the Recycling Dilemma," *Euromoney* (January 1975), pp. 28-29.

Hewson, John. "The Oil Crisis and World Financial Policies," *Euromoney* (January 1975), pp. 36-37.

Hunter, Robert E. *The Energy Crisis and U.S. Foreign Policy*. Washington: Overseas Development Council development paper 14, August 1973.

Johnson, Christopher. "Manila—The Olympic Games of International Finance," *The Banker*, Vol. 126 (November 1976), 1211-14.

Kleinman, D. "Oil Money and the Third World," *The Banker*, Vol. 124 (Summer 1974), p. 1061.

Kohjima, Sachio. "The OPEC Investment Managers Look to Their Pensions," *Euromoney* (July 1976), p. 72.

Levy, Walter J. "World Oil Cooperation or International Chaos," *Foreign Affairs*, Vol. 52 (July 1974), pp. 690-713.

Lomax, D. "Oil and International Debt." *The Banker*, Vol. 124 (April 1974), pp. 325-28.

Makdisi, Somer A. "The Role of the Oil Producers in the World Monetary System," *Euromoney* (March 1975), pp. 11-12.

Maynard, G. "Recycling of Oil Revenues," *The Banker*, Vol. 125 (January 1975), p. 39.

Mendelsohn, M. S. "Those Oil Deficits—Pick a Number," *Euromoney* (April 1974), pp. 22-23.

_____. "Oil and Payments: Who's Hitting Whom?" *Euromoney* (November 1973), p. 9.

Mikdashi, Zuhayr. "Cooperation among Oil Exporting Countries with Special Reference to Arab Countries: A Political Economy Analysis," *International Organization*, Vol. 28 (Winter 1974), pp. 1-30.

Moore, Alan. "Oil Money Now Flows in More Carefully Constructed Channels," *Euromoney* (September 1976), pp. 78-80.

Nashashibi, Hikmat Sh. "Surplus OPEC Funds Switch into the Euromarkets," *Euromoney* (October 1975), pp. 54-55.

_____. "Other Ways to Recycle Oil Surpluses," *Euromoney* (August 1974), p. 49.

Pollack, G. "The Economic Consequences of the Energy Crisis," *Foreign Affairs*, Vol. 52 (April 1974), pp. 452-71.

Smith, Dan. "The World Energy Situation Now—and the Implications for Oil Prices," *Euromoney* (June 1975), p. 36.

Truffert, Yves. "Some Reflections on the Oil Dollar Surpluses," *Euromoney* (May 1975), p. 43.

_____. "The Pattern of Arab Investment," *Euromoney* (September 1976), p. 76.

Williams, Maurice J. "The Aid Programs of the OPEC Countries," *Foreign Affairs*, Vol. 54 (January 1976), pp. 308-24.

Addresses and Speeches

Speech by His Excellency Shaikh Ali A. Alireza, Ambassador of Saudi Arabia to the United States, at National Foreign Trade Council Meeting, New York City, November 16, 1976.

Remarks of Mr. Fuji Matsumuro, Minister (Financial), Embassy of Japan, Washington, D.C., before the Balance of Payments Session, 63d National Foreign Trade Convention, New York City, November 16, 1976.

"Foreign Exchange and the Balance of Payments Position of the United States," Presented by John D. Wilson, Senior Vice President, The Chase Manhattan Bank, to the Balance of Payments Session, 63d National Foreign Trade Convention, New York City, November 16, 1976.

"World Trade and Investment in an Era of Rising International Debt," Speech by Richard D. Hill, Chairman of the Board, The First National Bank of Boston, at the 63d National Foreign Trade Convention, New York City, November 15, 1976.

Testimony of William P. Tavoulareas, President of Mobil Oil Corporation, before the Energy Subcommittee of the Joint Economic Committee of the Congress of the United States, June 2, 1976.

Remarks by Herman J. Schmidt, Vice Chairman, Mobil Oil Corporation, at Los Angeles Society of Financial Analysts, June 22, 1976.

"America's Widening Energy Gap," Remarks by A. E. Murray, President of U.S. Marketing and Refining Division of Mobil Oil Corporation, at the 74th Annual Meeting of the American Automobile Association, San Francisco, September 29, 1976.

Opening Remarks of Herman J. Schmidt, Vice Chairman, Mobil Oil Corporation, at the Symposium on Energy, 28th Annual Convention of the National Business Aircraft Association, New Orleans, October 29, 1975.

"Business Climate in the Middle East," Speech by Najeeb E. Halaby, President of Halaby International Corporation, at Overseas Automotive Club, New York City, October 9, 1975.

Opening Address by Gumersindo Rodriguez, Governor of the Bank for Venezuela at IMF–World Bank Joint Session, Washington, September 1, 1975.

"National Energy Needs," an address by Rawleigh Warner Jr., Chairman, Mobil Oil Corporation, Lubin Lectures, Pace University, New York City, February 26, 1974.

"The Eurocurrency Markets and Some New Roles for Banks in Recycling Petrodollars," Speech delivered by Carlos M. Canal, Jr., Executive Vice President, International Banking Department, Bankers Trust Company, to a conference sponsored by Institutional Investor Magazine, New York City, December 9, 1974.

"International Finance: An Arab Point of View," address by Abdlatif Y. Al-Hamad, Director-General of the Kuwait Fund for Arab Economic Development, at the Bankers Trust Company Anniversary Conference held in London, September 1974.

Index

About the Author

Anindya K. Bhattacharya is assistant professor of international business at the School of Business Administration of Adelphi University in New York. He is the author of *Foreign Trade and International Development* and *The Asian Dollar Market: International Offshore Financing.* His articles have appeared in *The Banker, Euromoney, The Bankers Magazine, Foreign Service Journal, International Organization, The C.F.A. Digest,* and *The Money Manager.*

Dr. Bhattacharya is a graduate of Cambridge University, England, and received the Ph.D. from Columbia University in New York.